Global Waste Management

Global Waste Management

Tom Dixon

STATES
ACADEMIC PRESS
www.statesacademicpress.com

Published by States Academic Press,
109 South 5th Street,
Brooklyn, NY 11249, USA
www.statesacademicpress.com

Global Waste Management
Tom Dixon

International Standard Book Number: 978-1-63989-242-6 (Hardback)

Cataloging-in-Publication Data

Global waste management / Tom Dixon.
p. cm.
Includes bibliographical references and index.
ISBN 978-1-63989-242-6
1. Refuse and refuse disposal. 2. Salvage (Waste, etc.). 3. Recycling (Waste, etc.).
4. Waste products. I. Dixon, Tom.
TD791 .G56 2022
363.728--dc23

TABLE OF CONTENTS

PREFACE

The different activities involved in the management of waste from its origin to its final disposal are studied under the subject of waste management. It involves processes related to collecting, transporting, treating and disposing of waste products, along with their monitoring and regulation. A few of the commonly used disposal methods are landfills and incineration. Recycling and reusing are also some of the important aspects of waste management. Commonly recycled materials include aluminum, copper, steel, rubber, glass, polyethylene and paper. There are various reusing techniques such as energy recovery, biological reprocessing, pyrolysis, energy recovery, resource recovery and sustainability. The transportation of waste materials is governed by the national laws of a country, but the trans-boundary movement of waste materials is often a subject of different international treaties. This book is compiled in such a manner, that it will provide in-depth knowledge about the theory and practice of global waste management. It elucidates new techniques and their applications in a multidisciplinary approach. This book will serve as a reference to a broad spectrum of readers.

To facilitate a deeper understanding of the contents of this book a short introduction of every chapter is written below:

Chapter 1- Any unwanted substance which is of no use and is discarded after primary use is waste. Some of the concepts studied in relation to waste are waste management, water characterization, waste processing, waste minimization, waste hierarchy, etc. This chapter serves as an introduction to the field of waste management.

Chapter 2- There are different types of waste such as hazardous waste, radioactive waste, e-waste, biomedical waste, solid waste and plastic waste. This chapter has been carefully written to provide an easy understanding of these types of wastes.

Chapter 3- Any garbage that has been discarded by the public falls under the category of solid waste. This chapter encompasses processes and techniques related to solid waste collection such as collection components, waste collection, collection operation, waste collection system design, storage, record keeping, etc. These topics are crucial for a complete understanding of the subject.

Chapter 4- Waste disposal refers to the various activities and actions necessary for the management of waste from its origin to final disposal. Some of the practices elaborated in this chapter related to waste disposal are disposal options and selection criteria, landfill, sanitary landfill, landfill gas emission, leachate formation, landfill operation issues, etc. This chapter discusses in detail the theories related to waste disposal.

Chapter 5- The process of conversion of waste materials into new and useful objects is known as recycling. This chapter explains some aspects of recycling such as plastic recycling, computer recycling, ship breaking, ferrous metal recycling and precycling in an easy-to-understand manner. The topics elaborated in this chapter will help in gaining a better perspective about recycling.

I owe the completion of this book to the never-ending support of my family, who supported me throughout the project.

Tom Dixon

Chapter 1

Waste Management: An Introduction

Any unwanted substance which is of no use and is discarded after primary use is waste. Some of the concepts studied in relation to waste are waste management, water characterization, waste processing, waste minimization, waste hierarchy, etc. This chapter serves as an introduction to the field of waste management.

Waste

Waste is defined as any unwanted or unusable substance that is discarded after primary use. Wastes are normally generated as a result of human and animal activities. Urbanization and rapid advancements in industrialization has led to an increase in the production and consumption processes resulting in the generation of wastes from various sectors that include agricultural, commercial, domestic, industrial, institutional, social and from community activities. Over time, these wastes accumulate and can have real impacts on the health and the environment. Waste management is intended to reduce adverse effects of waste on health, the environment or aesthetics. So to carry out efficient management of wastes, knowledge about the source of wastes, its types and classification must be known.

Classification of Wastes

Wastes are commonly classified based on the physical state as solid wastes, liquid wastes and gaseous wastes.

Solid Wastes

Solid wastes are any discarded or abandoned materials that can be solid, liquid, semi-solid or containerized gaseous material discarded by the human society. These include urban wastes, agricultural wastes, biomedical wastes and radioactive wastes. The term refuse is also used for solid wastes.

Examples of solid wastes include waste tires, septage, scrap metal, latex paints, furniture and toys, garbage, appliances and vehicles, oil and anti-freeze, empty aerosol cans, paint cans and compressed gas cylinders, construction and demolition debris, asbestos, plastics, styrofoam containers, bottles etc.

Liquid Wastes

Liquid wastes can be defined as liquids/fluids that are generated from washing, flushing or manufacturing processes of the industries. They are also called as sewage. The most common practice of disposing liquid waste is to discharge it in ground or rivers and other water bodies without treatment.

Examples: Domestic washings, chemicals, oils, waste water from ponds, Wastewater from manufacturing industries, manure, waste oil, fats, oils or grease (FOG), used oil, and hazardous household liquids.

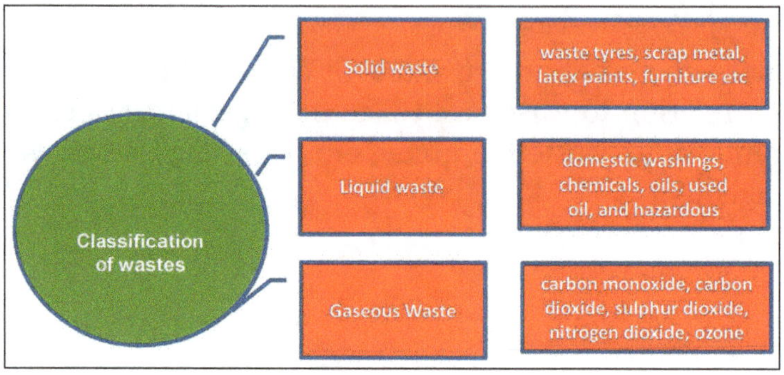

Figure: Classification of waste based on their physical state.

Gaseous Wastes

It is a waste product released in the form of gases from automobiles, factories, industries, burning of fossil fuels etc and gets mixed in the atmosphere. These gases include carbon monoxide, carbon dioxide, sulphur dioxide, nitrogen dioxide, ozone and methane etc.

Classification of Solid Wastes

It is mandatory to classify solid wastes into groups that pose similar risks to the environment and human health for safe disposal. According to the modern systems of waste management, solid wastes are classified based on their source, type, properties and its effect on human health and environment.

Source Based Classification

Wastes are produced from different sources and are categorized as follows:

- Municipal solid waste (MSW): Municipal solid waste commonly referred to as trash, garbage or refuse comprises of street wastes, dead animals, market wastes, abandoned vehicles, household garbage, rubbish, construction and demolition debris, sanitation residue, packaging materials, trade refuges etc. They are collected from residential houses, markets, streets and other places mostly from urban areas and disposed of by municipal bodies. The proportion of different constituents of municipal wastes vary from place to place and season to season depending on the food habits, life style, standard of living and extent of commercial and industrial activities in the area. Municipal wastes their contents and source are illustrated in table given below. Municipal solid wastes are further categorized based on their physical, chemical and biological properties.
- Industrial wastes: Wastes generated during industrial activities such as manufacturing

and processing involved in chemical plants, paint industry, cement factories, metallurgical plants, thermal power plants, petroleum, coal, gas, sanitary, textile, food processing and paper industry are referred to as industrial wastes. Some examples of industrial wastes are chemical solvents, paints, sandpaper, paper products, industrial by-products, metals, and radioactive wastes. Industrial solid wastes are further classified as hazardous and non-hazardous wastes.

- Institutional/Commercial wastes: Solid wastes originating from administrative, educational and public buildings such as offices, schools, colleges, hospitals, government centres, prisons and other commercial establishments like wholesale and retail stores, restaurants, hotels, markets, warehouses. Paper, cardboard, plastics, wood, food wastes, glass, metals, special wastes, hazardous wastes are the examples of industrial and commercial wastes.
- Agricultural wastes: Agriculture wastes include both natural (organic) and non-natural wastes generated through farming activities. These activities include but are not limited to dairy farming, horticulture, seed growing, livestock breeding, grazing land, market gardens, nursery plots, and even woodlands. Some of agricultural wastes include spoiled food grains, vegetables, animal and plant wastes, litter, pesticides, fertilizers etc. Other agricultural wastes are produced from agricultural products processing industries like sugarcane factories, tobacco processing units, slaughter houses, livestock, poultry etc. Agricultural wastes are mostly biodegradable but few wastes like pesticide and fertilizers are toxic. When discharged to the environment, agricultural wastes can be both beneficial and detrimental to living matter.

Table: Types of wastes with examples.

Properties	Type of wastes
Biodegradable waste	Food and kitchen waste, green waste, paper.
Recyclable materials	Paper, cardboard, glass, bottles, jars, tin cans, aluminium cans, foil, metals, certain plastics, fabrics, clothes, tyres, batteries.
Inert waste	Construction and demolition waste, dirt, rocks, debri
Electrical and electronic waste (WEEE)	Electrical appliances, light bulbs, washing machines, TVs, computers, screens, mobile phones, alarm clocks, watches etc.
Composite wastes	Waste clothing, tetra Packs, waste plastics such as toys
Hazardous waste	Paints, chemicals, tyres, batteries, light bulbs, electrical appliances, fluorescent lamps, aerosol spray cans, fertilizers.
Toxic waste	Pesticides, herbicides, and fungicides.
Biomedical waste	Expired pharmaceutical drugs.

Biomedical Wastes

Wastes produced from hospitals, medical centres and nursing homes are called biomedical wastes. Hospital wastes are generated during diagnosis, treatment or immunisation of human beings/animals or in research activities in these fields or in the production/testing of biologicals. These wastes are highly infectious and may pose severe threat if not managed properly. Biomedical wastes may be solid or liquid wastes that includes discarded blood, sharps, soiled wastes, disposables, anatomical wastes, cultures, discarded medicines, chemical wastes etc.

Type Based Classification

- Garbage: Garbage wastes mean and include animal and vegetable wastes generated from kitchen, cooking, serving of foods, slaughter houses, market refuse. These wastes contain putrescible organic matter which produces strong odour. They attract rats, vermin, flies and other insects and so they requires immediate attention in handling and disposal.
- Rubbish: Solid wastes arising as a result of households, commercial and institutional activities excluding garbage and ashes is termed as rubbish. They are categorized into combustible and non-combustible wastes. Combustible wastes consists of all rubbish and refuse that can be incinerated to flames at 1400-1500°F which includes leaves, plants, clothes, paper, leather, rubber, grasses. Non-combustible wastes are characterized as wastes that cannot be incinerated to flames at 1400-1500°F and that includes glass, metals, plastic materials, stones, auto parts etc.
- Bulk waste: Bulky household wastes consists of household furniture; appliances such as stoves, washing machines and refrigerators; mattresses and springs, rugs, TV sets, water heaters, tires, lawn mowers, auto parts, tree and brush debris etc.
- Commercial bulky wastes include packaging and containers such as cardboard, wood boxes, fiber, plastic and steel drums, loose and bundled paper, bundles of textiles and plastics, wires, furniture and equipment etc.
- Industrial bulky waste includes crates, cartons; steel, fiber and plastic drums; bales and rolls of paper, plastics, and textiles; miscellaneous metal items etc.
- Ashes: Ashes are defined as fine powdery residues, cinders and clinkers arising from the burning of wood, coal, charcoal, coke and other combustible materials during cooking and heating in houses, institutions and outer industrial establishments.

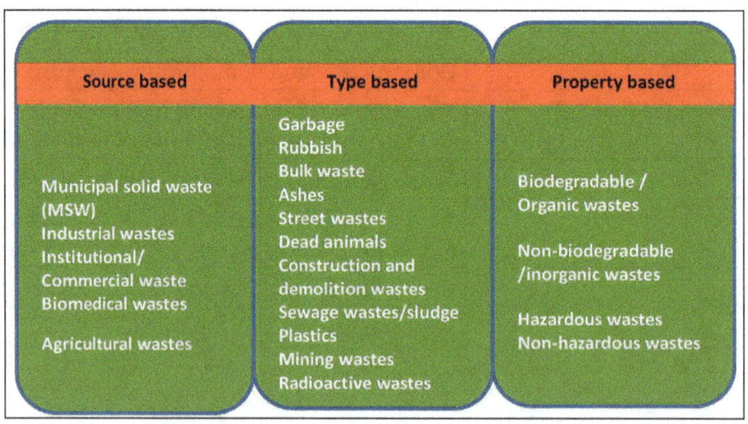

Figure: Classification solid waste based on source of generation, type and physio-chemical property.

- Street wastes: Wastes comprising of leaves, dirt, dust litter, paper, plastics and other vegetable matter collected from streets, walkways, alleys, parks, beaches and vacant lots are termed as street wastes.
- Dead animals: Animals that die naturally or accidentally comprises of dead animal wastes.

They are categorized as small animals such as cats, dogs, rats, rabbits, poultry and large animals like horses, cows, goats, Sheep etc. However, animal carcasses and animal parts from slaughterhouses are excluded from dead animal wastes and considered as industrial wastes. These dead animal wastes are putrescible and attract flies and other vermin posing severe health risk to the environment. Hence they must be collected and disposed off promptly.

- Construction and demolition wastes: Construction and demolition wastes are the waste materials generated in large amounts during the construction, refurbishment, repair and demolition of houses, commercial buildings, roads and other structures. They consists of earth, stones, concrete, bricks, lumber, steel, roofing materials, plumbing materials, heating systems and electrical wires.

- Sewage wastes/sludge: Settled solid components, residual or semi-solid materials that are discharged from sewage treatment plants and septic tanks are classified as sewage waste. The raw and the treated sewage consist of organic and inorganic fraction. The sewage wastes contains putrescible organic matter which may contain pathogens and so must be disposed off without delay.

- Plastics: Plastic due to their versatile property of being light, durable, easy to mould and economical has invaded almost all sectors of the economy. Likewise, they are generated as wastes from almost all sectors that includes agriculture, construction, consumer goods, household, health care, hotel and catering, packaging, telecommunications, air and travel industries. Some of the plastic wastes include carry bags, bottles, plates, spoons, glasses, gloves, boxes, syringes, catheter tubes, surgical items etc. Plastics due to its non-biodegradable nature are now considered a serious threat to the environmental and health.

- Mining wastes: Mining wastes are generated from extractive operations of mineral resources. They include materials such as topsoil, overburden and waste rock that must be removed to gain access to the mineral resource. Also, other waste material like slags, mine water, mine tailings, water treatment sludge and gaseous wastes etc are released during or after processing of mineral ores. Some of these wastes are inert and are not considered as threat to the environment. However, other fractions, generated by the non-ferrous metal mining industry contains large quantities of dangerous substances such as heavy metals. These metals and metal compounds after extraction and subsequent mineral processing, tend to become chemically more available resulting in the generation of acid or alkaline drainage. Therefore, mine wastes requires to be carefully characterized to prevent and minimize air, water, and soil contamination.

- Radioactive wastes: Radioactive wastes are hazardous, by-products of nuclear reactions. They pose severe threat to human life and environment. Radioactive wastes decays over time ranging from a few days for highly radioactive isotopes to millions of years for slightly radioactive ones. Hence, these wastes have to be isolated and confined at appropriate disposal facilities for it to completely decay. The sources of radioactive wastes are from mining of radioactive substances, atomic explosion, nuclear fuel cycle, nuclear weapons reprocessing, medical and industrial wastes etc.

Property Based Classification

Solid wastes are also classified based on their biological and chemical property:

- Biodegradable/Organic wastes: Biodegradable wastes are those that can be decomposed by the natural processes such as composting, aerobic/ anaerobic digestion and converted into the elemental form like carbon dioxide, methane, water or simple organic molecules. Some of the biodegradable wastes include municipal solid wastes (green waste, food waste, paper waste, biodegradable plastics, human and animal wastes, sewage, sludge, slaughter house wastes etc).
- Non-biodegradable/inorganic wastes: Non-biodegradable wastes are those that cannot be decomposed and remain as such in the environment indefinitely. They are persistent and threaten to overwhelm landfills and create disposal problems creating environmental concern. As non-biodegradable wastes cannot be decomposed, recycling is the ideal option for managing it. Example of non-biodegradable wastes includes plastics, nuclear wastes, glass, rubber tyres, styrofoam, fiberglass and metals.
- Hazardous wastes: Hazardous waste is defined as chemical material that can no longer be used for its intended purpose and is known to be harmful or potentially harmful to plants, animals and human health or to the environment. Hazardous wastes may be in the form of solids, liquids, sludge's or gases. In some cases, although the active agents may be liquid or gaseous, they are classified as solid waste because they are confined in solid containers. They are generated primarily by chemical production, manufacturing and other industrial activities. The hazardous waste materials may be toxic, reactive, ignitable, explosive, corrosive, infectious or radioactive. If improperly handled, they can cause substantial harm to human health and to the environment. So good management practice should ensure that hazardous wastes are collected, stored, transported and disposed off separately, to render them innocuous. Some of the important hazardous wastes are lead, mercury, cadmium, chromium, many drugs, leather, pesticides, dye, rubber, solvents, paints and effluents from different industries.
- Non-hazardous wastes: Non-hazardous wastes are defined as substances safe to use commercially, industrially, agriculturally or economically. Some of the non-hazardous wastes produced are from the food processing plants, cotton mills, paper mills, textile mills and sugarcane industries. Other non-hazardous wastes includes paint, oil, antifreeze, buffers, salts etc.

Waste Management

Waste management is the collection, transport, processing, recycling and/or disposal of waste materials produced by human activity. Waste management reduces the effect of waste on the environment, health, and so on. It can also help reuse or recycle resources, such as; paper, cans, glass, and so on. There are various type of waste management that include the disposal of solid, liquid, gaseous or hazardous substances.

Characteristics of Waste Management

Waste management practices differ for developed and developing nations, for urban and rural areas, and for residential and industrial, producers. Management for non-hazardous residential and institutional waste in metropolitan areas is usually the responsibility of local government authorities, while management for non-hazardous commercial and industrial waste is usually the responsibility of the generator. When recycling began to be recognized as essential for both environmental and resource management reasons, recycling rates for household wastes in most developed countries in the 1980s were in the low single figures by percent. Modern western waste management systems have rebuilt recycling rates over the last 20 years. Modern waste management systems, which many developing country cities aspire to, are all characterized by high recycling rates of clean, source separated materials.

Most technologies for waste management are mature and have been successfully implemented for decades in many countries. In Turkey, Solid Waste Control Regulation came into force in order to manage solid waste in 1991. The regulation has been continuously updated. Considering the Directive of Waste Frame EEC and the other directives of the European Union as part of the requirements of the integration process into the EU, the Department has started the preparations for the Regulation of Waste Frame and the draft of the Regulation has been presented for consideration. Finally, the waste management is conducted according to the Turkish Solid Waste Management that was adopted in 2008. In the Regulation, the procedures about the classification of the wastes, collection, transportation, and temporary storage of the wastes within the institutions and the transportation of them to the final disposal area were explained. In the last part of the Regulation, the method of the disposal of the wastes was mentioned. A waste management concept including the following goals:

- Reduction of total amount of waste by reduction and recycling of refuse.
- Recycling and re-introduction of suitable groups of substances into production cycles as secondary raw material or energy carrier.
- Re-introduction of biological waste into the natural cycle.
- Best-possible reduction of residual waste quantities, which are to be disposed on "suitable" landfills.
- Flexible concept concerning fluctuations in waste quantities and the composition of domestic waste. New developments in the field of waste management must be included into the system.

Large quantities of this waste cannot be eliminated. However, the environmental impact can be reduced by making more sustainable use of this waste. This is known as the "Waste Hierarchy". The waste hierarchy refers to the "3 Rs" reduce, reuse and recycle, which classify waste management strategies according to their desirability in terms of waste minimization. The hierarchy of disposal options, which categorizes environmental impacts into six levels, from low to high; namely, reduce, reuse, recycle, compost, incinerate and landfill. The waste hierarchy remains the cornerstone of most waste minimization strategies. The aim of the waste hierarchy is to extract the maximum practical benefits from products and to generate the minimum amount of waste.

Waste collection methods vary widely between different countries and regions. Domestic waste collection services are often provided by local government authorities, or by private industry. Some areas, especially those in less developed countries, do not have a formal waste-collection system.

Waste Management Methods

Classification and assignment of different waste materials to classes cannot be easily achieved by conventional parameters. Nowadays not all the wastes are classified under the same classical category, requiring the same procedure of elimination. Especially in the industries in which the types of wastes are so varied it is very difficult and impractical to set up and operate a different management system for different types of wastes. Thus has emerged the necessity of a, management system which would eliminate all types of wastes. Table 1 has presented a selection of promising waste treatment processes.

Disposal Methods

An integrated approach is required in an attempt to manage such large quantities of a diverse, contaminated mixture of wastes in an energy efficient and environmentally benign manner. This would require examining critically various steps in the life of the wastes such as the raw materials for their manufacture, the manufacturing processes, design and fabrication of the finished products, possible reuse of those items, and the proper disposal of the wastes, in totality. Such an integrated waste management concept comprises: (i) source reduction; (ii) reuse; (iii) recycling; (iv) landfill and gas-to energy (v) waste-to energy conversion.

Table: Selected waste treatment processes.

Waste stream		Waste disposal methods
Combustible wastes		Roaster incineration. Fluid bed incineration. Pyrolysis-incineration. Pyrolysis-gasification. Separation-composting-incineration. (Wet and dry) separation-digesting-incineration. Separation-digesting-pyrolysis. Separation-digesting-gasification. Separation-digesting-incineration in a cement plant. Selective separation-incineration.
Non-combustible wastes Landfill		Landfill
Partially combustible waste streams.	Wood	Pyrolysis and co-incineration in a coal power plant. Pyrolysis and co-incineration in a powdered coal. power plant Incineration in a fluid bed furnace Gasification.
	Plastics	Gasification Feedstock Recycling
	Organic wastes	Composting Anaerobic Digestion

Landfill

A landfill is not a normal environmental condition, though, nor is it intended to be. Instead, a landfill is more like a tightly sealed storage container. A landfill is designed to inhibit degradation to protect the environment from harmful contamination. Deprived of air and water, even organic wastes like paper and grass clippings degrade very slowly in a landfill. Municipal Solid Waste Landfills (MSWLs) are places where heterogeneous mixtures of municipal and industrial wastes are disposed. Mechanical properties of MSW such as strength and compressibility depend on composition of waste, mechanical properties of components of the material deposited, water content and decomposition.

Disposing of waste in a landfill involves burying waste to dispose of it, and this remains a common practice in most countries. Landfills were often established in disused quarries, mining voids or borrow pits. A properly-designed and well-managed landfill can be a hygienic and relatively inexpensive method of disposing of waste materials. Older, poorly designed or poorly-managed landfills can create a number of adverse environmental impacts such as wind-blown litter, attraction of vermin (especially, noxious little animals or insects, collectively, as squirrels, rats, mice, flies, lice, bugs, etc), and generation of liquid leachate.

Landfill leachates contain a large number of compounds, some of which can be expected to create a threat to health and nature if released into the natural environment. For this purpose, Öman and Junestedt, have been studied screened samples from 12 Swedish municipal landfill sites for 400 parameters and compounds which expected to be present in landfill leachate. More than 90 organic and metal organic compounds and 50 inorganic elements were detected, some of which seem to have not been detected before. Compounds detected include halogenated aliphatic compounds, benzene and alkylated benzenes, phenol and alkylated phenols, ethoxylates, polycyclic aromatic compounds, phthalic esters, chlorinated benzenes, chlorinated phenols, PCB, chlorinated dioxins and chlorinated furans, bromated flameretardants, pesticides, organic tin, methyl mercury and heavy metals. The presence of this large number of hazardous compounds in landfill leachates should have a significant impact on future landfill risk assessments and the development of leachate treatment methods.

Today's landfills are very different from the open dumps of the past. For one thing, new landfills are situated where clay deposits and other land features act as natural buffers between the landfills and the surrounding environment. Second, the bottom and sides of modern landfills are lined with layers of clay or plastic to keep the liquid waste, called leachate, from escaping into the soil. Deposited waste is normally compacted to increase its density and stability, and covered to prevent attracting vermin.

Another common by product of landfills is gases. Landfill gas is a renewable energy source. Methane (CH_4) and carbon dioxide (CO_2) are major gases, forming from the anaerobic degradation of degradable domestic solid wastes. Landfill gases generally contain 45-60% CH_4 and 40-60% CO_2. These gases can create odor problems, kill surface vegetation, and is a greenhouse gas (GHG). Some of these compounds may cause corrosion or other damage to engines if they are not removed from the gas before utilization.

Landfills are an important source of atmospheric CH_4 due to methanogenic activity under anaerobic conditions. Landfill CH_4 recovery and optimized wastewater treatment can directly reduce GHG emissions. So far, studies and applications on upgrading landfill gas are relatively few, while the upgrading of biogases from sewage and biowaste digesters is rapidly increasing in some countries. For example, in Reykjavik (Iceland), landfill gas upgrading was started in 2000 with water scrubber technology. In 2003, East Kentucky Power Cooperative began recovering methane gas from three landfills. The utility uses the landfill gas to generate 8.8 MW of electricity, enough power 7500-8000 homes.

Landfilling is the last choice method in the hierarchy of waste management. However, even with sanitary landfilling being the most common method of solid waste management, there was no landfill gas energy recovery. With the scarcity of space for landfilling and due to its ever increasing cost, waste utilization has become an attractive alternative to disposal.

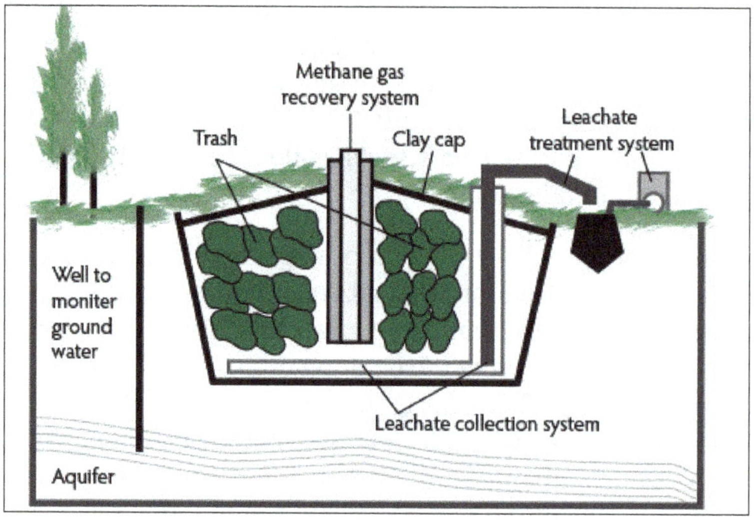

Figure: Modern landfill.

Biorenewable Gaseous Fuels

Liquid (bioethanol and biodiesel) and gaseous (biogas and bio-hydrogen) biofuels have become more attractive recently because of its environmental benefits. In recent years, many studies have investigated the economic and environmental impacts of the biofuels, especially bioethanol and biodiesel. Liquid biofuels are important because they replace petroleum fuels. There are many benefits for the environment, economy and consumers in using biofuels. Biofuels production costs can vary widely by feedstock, conversion process, scale of production and region. Biofuels production costs can vary widely by feedstock, conversion process, scale of production and region. On an energy basis, ethanol is currently more expensive to produce than gasoline in all regions considered.

Biogas can be obtained from several sources. It is obtained from decomposing organic material. Contents of domestic solid waste are given in Table. Biogas is composed by methane (CH_4), carbon dioxide (CO_2), air, ammonia, carbon monoxide, hydrogen, sulphur gases, nitrogen and oxygen. Among its components, methane is the most important one, particularly for the combustion process in vehicle engines. A typical analysis of raw landfill gas is given in Table. CH_4 and CO_2 make up around 90% of the gas volume produced. The main constituents of landfill gas are methane and carbon dioxide, both of which are major contributors to global warming. Because of the widely varying nature of the contents of landfill sites the constituents of landfill gases vary widely.

Landfill leachate treatment has received significant attention in recent years, especially in municipal areas. The generation of municipal solid wastes (MSW) has increased in parallel to rapid industrialization. Approximately 16% of all discarded MSW is incinerated; the remainder is disposed of in landfills. Effective management of these wastes has become a major social and environmental concern. Disposal of MSW in sanitary landfills is usually associated with soil, surface water and groundwater contamination when the landfill is not properly constructed. The flow rate and composition of leachate vary from site to site, seasonally at each site and depending on the age of the landfill. Young leachate normally contains high amounts of volatile fatty acids. MSW statistics and management practices including waste recovery and recycling initiatives have been evaluated. The organic MSW was chemically and biologically characterized, in order to study its behaviour during anaerobic digestion, and its pH, biogas production, alkalinity, and volatile fatty acid production was determined by Plaza et al. Anaerobic digestion of the organic food fraction of MSW, on its own or co-digested with primary sewage sludge, produces high quality biogas, suitable as renewable energy. The processing of MSW (i. e. landfill, incineration, aerobic composting) secures many advantages and limitations. The greenhouse gas emissions can be reduced by the uncontrolled releasing of methane from improperly disposed organic waste in a large landfill.

Table: Contents of domestic solid waste (wt% of total).

Component	Lower limit	Upper limit
Paper waste	33.2	50.7
Food waste	18.2	21.2
Plastic matter	7.8	11.2
Metal	7.3	10.5
Glass	8.6	10.2
Textile	2.0	2.8
Wood	1.8	2.9
Leather and rubber	0.6	1.0
Miscellaneous	1.2	1.8

Table: Typical analysis of raw landfill gas.

Component	Chemical formula	Content
Methane	CH_4	40-60 (% by vol.)
Carbon dioxide	CO_2	20-40 (% by vol.)
Nitrogen	N_2	2-20 (% by vol.)
Oxygen	O_2	<1 (% by vol.)
Heavier hydrocarbons	C_nH_{2n+2}	<1 (% by vol.)
Hydrogen sulfide	H_2S	40-100 ppm
Complex organics	-	1000-2000 ppm

Decomposition in landfills occurs in a series of stages, each of which is characterized by the increase or decrease of specific bacterial populations and the formation and utilization of certain metabolic products. The first stage of decomposition, which usually lasts less than a week, is characterized by the removal of oxygen from the waste by aerobic bacteria. In the second stage, which has been termed the anaerobic acid stage, a diverse population of hydrolytic and fermentative bacteria hydrolyzes polymers, such as cellulose, hemicellulose, proteins, and lipids, into soluble sugars, amino acids, long-chain carboxylic acids, and glycerol. Figure shows the behavior of biogas production with time, in terms of the biogas components. Figure indicates that the economic exploitation of CH_4 is worthwhile after one year from the start of the landfill operation. The main components of landfill gas are by products of the decomposition of organic material, usually in the form of domestic waste, by the action of naturally occurring bacteria under anaerobic conditions.

Methods developed for treatment of landfill leachates can be classified as physical, chemical and biological which are usually used in combinations in order to improve the treatment efficiency. Biological treatment methods used for the leachate treatment can be classified as aerobic, anaerobic and anoxic processes which are widely used for the removal of biodegradable compounds. Biological treatment of landfill leachate usually results in low nutrient removals because of high chemical oxygen demand (COD), high ammonium-N content and the presence of toxic compounds such as heavy metals. Landfill leachate obtained from the solid waste landfill area contained high COD and ammonium ions which resulted in low COD and ammonium removals by direct biological treatment. Several anaerobic and aerobic treatment systems have been studied in landfill leachate. Leachates contain non-biodegradable substrates which are not removed by biological treatment alone and an increase of leachate input may cause reduction in substrate removal. Raw landfill leachate was subjected to pre-treatment by coagulation-flocculation and air stripping of ammonia before biological treatment. In order to improve biological treatability of the leachate, coagulation-flocculation and air stripping of ammonia were used as pre-treatment. Natural zeolite and bentonite can be utilized as a novel landfill liner material.

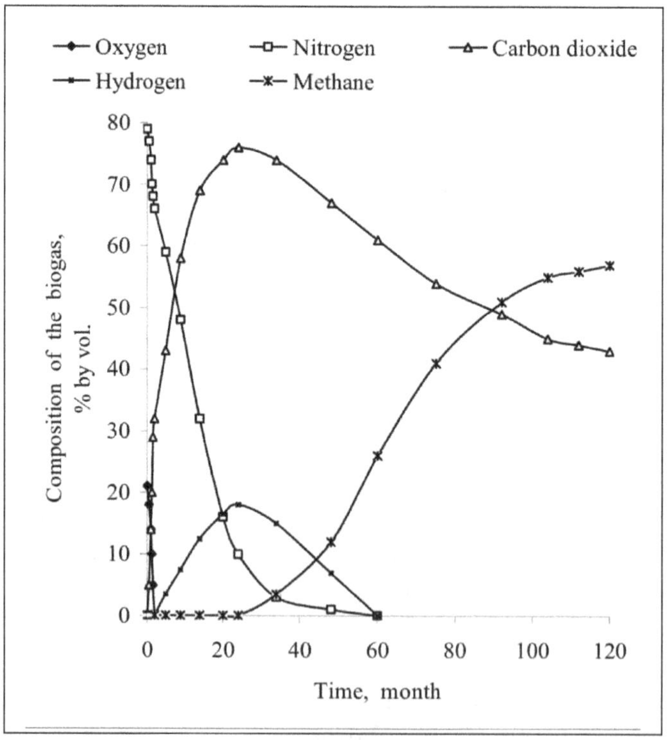

Figure: Production of biogas components with time in landfill.

Biogas Technology

Anaerobic digestion (AD) is the conversion of organic material directly to a gas, termed biogas, a mixture of mainly methane and carbon dioxide with small quantities of other gases such as hydrogen sulfide. Methane is the major component of the biogas used in many homes for cooking and heating. The biogas has a chemical composition close to that of natural gas, the biodigester, or a biogas plant, is a physical structure used to provide an anaerobic condition which stimulates various chemical and microbiological reactions resulting in the decomposition of input slurries and the production of biogas—mainly methane.

Biogas can be used after appropriate gas clean up as a fuel for engines, gas turbines, fuel cells, boilers, industrial heaters, other processes, or for the manufacturing of chemicals. Before landfilling, treatment or stabilization of biodegradable materials can be accomplished by a combination of anaerobic digestion followed by aerobic composting.

The same types of anaerobic bacteria that produced natural gas also produce methane today. Anaerobic bacteria are some of the oldest forms of life on earth. They evolved before the photosynthesis of green plants released large quantities of oxygen into the atmosphere. Anaerobic bacteria break down or digest organic material in the absence of oxygen and produce biogas as a waste product.

The first methane digester plant was built at a leper colony in Bombay, India, in 1859. Most of the biogas plants utilize animal dung or sewage. Schematic of biogas plant utilizing cow dung is illustrated in Fig. Anaerobic digestion is a commercially proven technology and is widely used for treating high moisture content organic wastes including +80%–90% moisture. Biogas can be used directly in spark ignition gas engines (SIGEs) and gas turbines. Used as a fuel in SIGE to produce electricity only, the overall conversion efficiency from biomass to electricity is about 10%–16%.

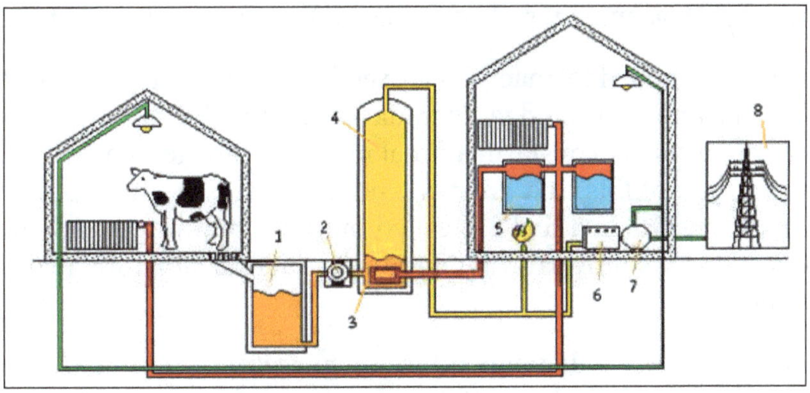

Figure: Schematic of biogas plant utilizing cow dung: 1. Compost storage, 2. Pump, 3. Internal heater, 4. Digester, 5. Combustor, 6-8. Power generator.

Aerobic Conversion Processes

Aerobic conversion includes most commercial composting and activated sludge wastewater treatment processes. Aerobic conversion uses air or oxygen to support the metabolism of the aerobic microorganisms degrading the substrate. Nutritional considerations are also important to the proper functioning of aerobic processes. Aerobic processes operate at much higher reaction rates than anaerobic processes and produce more cell mass, but generally do not produce useful fuel gases. Aerobic decomposition can occur from as low as near freezing to about 344 K.

Respiration refers to those biochemical processes in which organisms oxidize organic matter and extract the stored chemical energy needed for growth and reproduction. Respiration patterns may be subdivided into two major groups, based on the nature of the ultimate election acceptor.

Although alternative pathways exist for the oxidation of various organic substrates, it is convenient to consider only the degradation of glucose. The breakdown of glucose is via the Embden-Meyer-of-Parnas glycolytic pathway which yields 2 moles each of pyruvate, ATP, and reduced nicotinamide adenine dinucleotide (NAD) per mole of glucose.

Under aerobic conditions, the pyruvate is oxidized to CO_2 and H_2O via the tricarboxylic acid or Krebs cycle and the electron transport system. The net yield for glycolysis followed by complete oxidation is 38 moles ATP per mole glucose, although there is evidence that the yield for bacteria is 16 moles ATP per mole glucose. Thus, 673 kcal are liberated per mole glucose, much of which is stored as ATP.

Anaerobic Conversion Processes

Anaerobic digestion (AD) is a bacterial fermentation process that is sometimes employed in wastewater treatment for sludge degradation and stabilization. This is also the principal process occurring in the decomposition of food wastes and other biomass in landfills. The AD operates without free oxygen and results in a fuel gas called biogas, containing mostly CH_4 and CO_2 but frequently carrying other substances such as moisture, hydrogen sulfide (H_2S), and particulate matter that are generally removed prior to use of the biogas. The AD is a biochemical process for converting biogenic solid waste into a stable, humus-like product. Aerobic conversion uses air or oxygen to support the metabolism of the aerobic microorganisms degrading the substrate. Aerobic conversion includes composting and activated sludge wastewater treatment processes. Composting produces useful materials, such as mulch, soil additives and amendments, and fertilizers.

Digestion is a term usually applied to anaerobic mixed bacterial culture systems employed in many wastewater treatment facilities for sludge degradation and stabilization. Anaerobic digestion is also becoming more widely used in on-farm animal manure management systems, and is the principal process occurring in landfills that creates landfill gas (LFG). Anaerobic digestion operates without free oxygen and results in a fuel gas called biogas containing mostly methane (CH_4) and carbon dioxide (CO_2), but frequently carrying impurities such as moisture, hydrogen sulfide (H_2S), and particulate matter.

The AD is known to occur over a wide temperature range from 10 to 71 °C. Anaerobic digestion requires attention to the nutritional needs and the maintenance of reasonable temperatures for the facultative and methanogenic bacteria degrading the waste substrates. The carbon/nitrogen (C/N) ratio of the feedstock is especially important. Biogas can be used after appropriate gas cleanup as a fuel for engines, gas turbines, fuel cells, boilers, industrial heaters, other processes, and the manufacturing of chemicals. Anaerobic digestion is also being explored as a route for direct conversion to hydrogen.

Cellulose and hemicelluloses can be hydrolyzed to simple sugars and amino acids that are consumed and transformed by the fermentive bacteria. The lignin is refractory to hydrolysis and generally exits the process undigested. In fact, lignin may be the most recalcitrant naturally produced organic chemical. Lignin polymers are cross-linked carbohydrate structures with molecular weights on the order of 10,000 atomic mass units. As such, lignin can bind with or encapsulate some cellulose making that cellulose unavailable to hydrolysis and digestion. Lignin degradation (or delignification of lignocellulosics) in nature is due principally to aerobic filamentous fungi that decompose the lignin in order to gain access to the cellulose and hemicelluloses.

For anaerobic systems, methane gas is an important product. Depending on the type and nature of the biological components, different yields can be obtained for different biodegradable wastes. For pure cellulose, for example, the biogas product is 50 % methane and 50 % carbon dioxide. Mixed waste feedstocks yield biogas with methane concentrations of 40-60 % (by volume). Fats and oils can yield biogas with 70 % methane content.

Anaerobic digestion functions over a wide temperature range from the so-called psychrophilic temperature near 283 K to extreme thermophilic temperatures above 344 K. The temperature of the reaction has a very strong influence on the anaerobic activity, but there are two optimal temperature ranges in which microbial activity and biogas production rate are highest, the so called mesophilic and thermophilic ranges. The mesophilic regime is associated with temperatures of about 308 K, the thermophilic regime of about 328 K. Operation at thermophilic temperature allows for shorter retention time and a higher biogas production rate, however, and maintaining the high temperature generally requires an outside heat source because anaerobic bacteria do not generate sufficient heat. Aerobic composting can achieve relatively high temperatures (up to 344 K) without heat addition because reaction rates for aerobic systems are much higher than those for anaerobic systems. If heat is not conducted away from the hot center of a compost pile, then thermochemical reactions can initiate which can lead to spontaneous combustion if sufficient oxygen reaches the hot areas.

Managed compost operations use aeration to provide oxygen to the bacteria but also to transport heat out of the pile. The anaerobic digestion of lignocellulosic waste occurs in a three-step process often termed hydrolysis, acetogenesis, and methanogenesis, as shown schematically in Figure. The molecular structure of the biodegradable portion of the waste that contains proteins and carbohydrates is first broken down through hydrolysis. The lipids are converted to volatile fatty acids and amino acids. Carbohydrates and proteins are hydrolyzed to sugars and amino acids. In acetogenesis, acid forming bacteria use these by-products to generate intermediary products such as propionate and butyrate. Further microbial action results in the degradation of these intermediary products into hydrogen and acetate. Methanogenic bacteria consume the hydrogen and acetate to produce methane and carbon dioxide.

Under anaerobic conditions, various pathways exist for pyruvate metabolisms which serve to reoxidize the reduced hydrogen carriers formed during glycolysis. The ultimate acceptor builds up as a waste product in the culture medium. The end products of the pathways are: (1) CO_2, ATP, and acetate; (2) CO_2 and ethanol; (3) H_2 and CO_2; (4) CO_2 and 2,3-butylene glycol; (5) CO_2, H_2, acetone, ATP, and butanol; (6) succinate; and (7) lactate. The pathway that occurs depends on the microorganism cultivated and the culture.

Biogas Processing

A methane digester system, commonly referred to as an AD is a device that promotes the decomposition of manure or digestion of the organics in manure to simple organics and gaseous biogas products. There are three types of continuous digesters: vertical tank systems, horizontal tank or plug-flow systems, and multiple tank systems. Proper design, operation, and maintenance of continuous digesters produce a steady and predictable supply of usable biogas.

Biogas, a clean and renewable form of energy could very well substitute (especially in the rural sector) for conventional sources of energy (fossil fuels, oil, etc.) which are causing ecological–environmental problems and at the same time depleting at a faster rate. Biogas is a modern form of bioenergy that is derived from the anaerobic digestion of organic matter, such as manure, sewage sludge, municipal solid waste, biodegradable waste and agricultural slurry under anaerobic conditions.

Anaerobic digestion of organic compounds is a complex process, involving several different types of microorganisms. This is the natural breakdown of organic matter, such as biomass, by bacterial populations in the absence of air into biogas, i.e., a mixture of methane (40-75% v/v) and carbon dioxide. The end products of anaerobic digestion are biogas and digestate, a moist solid which is normally dewatered to produce a liquid stream and a drier solid. During anaerobic digestion, typically 30-60% of the input solids are converted to biogas; by products consist of undigested fibre and various water-soluble substances.

The anaerobic digestion process occurs in the following four basic steps: (1) hydrolysis, (2) acidogenesis, (3) acetogenesis, and (4) methanogenesis. A simplified model of anaerobic digestion process, showing the main steps, is shown in Fig.

Incineration

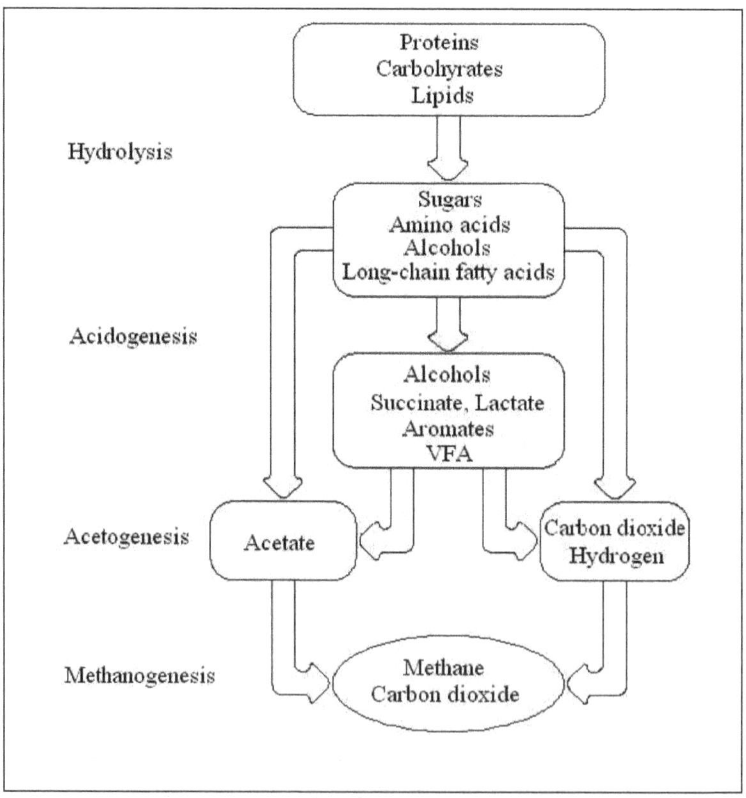

Figure: A simplified conversion processes in anaerobic digestion.

Incineration is a disposal method that involves combustion of waste material. Incineration and other high temperature waste treatment systems are sometimes described as "thermal treatment". Incinerators convert waste materials into heat, gas, steam, and ash. a solid waste incinerator is a type of facility which is designed, built, and operated at specified design conditions. A typical incinerator processes wastes that have been collected as input material, and achieves its goal, i.e., treatment of waste material and as secondary benefit recovers heat energy from the combustion process.

Incineration is carried out both on a small scale by individuals and on a large scale by industry. It is used to dispose of solid, liquid and gaseous waste. It is recognized as a practical method of disposing of certain hazardous waste materials (such as biological medical waste). Incineration is a controversial method of waste disposal, due to issues such as emission of gaseous pollutants. Solid waste incineration generates solid residues, such as bottom ash and air pollution control residues. Besides a high content of inorganic compounds, incineration residues also contain abundant carbon compounds deriving from incomplete combustion, unburned organic matter and carbon compounds formed during the incineration process. In addition, the incineration of fossil carbon results in minor emissions of CO_2.

Incineration is common in countries such as Japan where land is scarcer, as these facilities generally do not require as much area as landfills. In Japan, where open space is very limited for construction of waste management infrastructure, very high rates of both recycling and incineration are practiced and are expected to continue into the future.

Waste-to-energy (WTE) or energy-from-waste (EFW) is broad terms for facilities that burn waste in a furnace or boiler to generate heat, steam and/or electricity. Combustion in an incinerator is not always perfect and there have been concerns about micro-pollutants in gaseous emissions from incinerator stacks. Particular concern has focused on some very persistent organics such as Dioxins which may be created within the incinerator and which may have serious environmental consequences in the area immediately around the incinerator. Gullet et al. have reported the discovery of polychlorinated dibenzo-p-dioxins (PCDDs) and polychlorinated dibenzofurans (PCDFs) from waste combustion sources and the increased under-standing of the toxic effects of these chemicals have prompted extensive study over the last 10 or so years on how these trace compounds are formed in combustion systems.

Incineration for waste to energy (WTE) has been widely implemented in many European countries for decades. In 2002, EU WTE plants generated 41 million GJ of electrical energy and 110 million GJ of thermal energy. Incineration and industrial co-combustion for WTE provide significant renewable energy benefits and fossil fuel offsets. Currently, over 130 million tonnes of waste per year are incinerated at over 600 plants. Thermal processes with advanced emission controls are proven technology but more costly than controlled landfilling with landfill gas recovery; however, thermal processes may become more viable as energy prices increase.

Because landfills produce CH_4 for decades, incineration, composting and other strategies that reduce land filled waste are complementary mitigation measures to landfill gas recovery in the short- to medium-term.

Waste Characterization

Waste characterization means finding out how much paper, glass, food waste, etc. is discarded in your waste stream. Waste characterization information helps in planning how to reduce waste, set up recycling programs, and conserve money and resources.

Who Should use this Information?

Waste characterization information is designed for solid waste planning; however, anyone interested in the characteristics of the solid waste stream may find it useful. Local government planners, haulers, and recyclers can use this site's waste characterization tools to estimate the amount of certain materials in their waste stream.

Waste characterization data is collected by taking samples of waste and sorting it into material types like newspaper and aluminum cans, and weighing each type. Typically, samples are taken from trucks delivering waste to landfills and transfer stations from residential, commercial, and self-haul sources. In some cases, samples are taken from individual businesses to develop waste composition data for specific types of businesses (often called a generator-based study).

Let us take the case of Municipal Solid Waste (MSW), this stream of waste has various components like organic waste, plastics, metal cans, paper, cardboard, glass, inert materials etc., Some of the constituents in this waste can be recycled, while some can be reduced and used and some others have to be only disposed. Thus, the action taken depends on the nature of the waste and its constituents. This is where the usefulness of waste characterization lies. Waste characterization is a method that is used to determine the types and proportion of materials that are being discarded in a waste stream. Waste characterization is essential in determining the management practices that are to be adopted to manage them. It is especially useful for policy makers and city planners who are involved in identifying and assigning landfill sites and setting up recycling programs.

Since the characteristic of waste determines the method of its management it is important and useful to know about the physical, chemical and biological nature of the waste. These characteristics vary depending on the source and type of solid waste and this in turn will affect the leachate and gas production from landfills.

Physical Characteristic of MSW

Coming to MSW the physical characteristic of the waste includes moisture content, waste particle size, waste density, temperature and pH as these affect the extent and rate of degradation of waste.

Moisture Content

The moisture content of solid wastes usually is expressed as the weight of moisture per unit weight of wet or dry material. In the wet-weight method of measurement, the moisture in a sample is expressed as a percentage of the wet weight of the material; in the dry-weight method, it is expressed as a percentage of the dry weight of the material. In equation form, the wet-weight moisture content is expressed as follows:

$$\text{Moisture content}(\%) = \left(\frac{a-b}{a}\right)100$$

where, a = initial weight of sample as delivered, b = weight of sample after drying.

Typically in MSW the moisture content varies between 15-40% and this further depends on the composition of the wastes, the season of the year, and the humidity and weather conditions.

Moisture content is a critical determinant in the economic feasibility of waste treatment by incineration since energy must be supplied for evaporation of this moisture. Moisture content also plays an important role in other processing methods such as composting and anaerobic digestion. Most micro-organisms including bacteria require a minimum of approximately 12% moisture for growth.

Density

Density of a waste is its mass per unit volume (kg/m^3). It is essential for the design of all elements of the solid waste management system from storage to transportation to final disposal. In high income countries, the collected waste is typically of low density as it contains more recyclables like cans, glasses etc. Here considerable benefit is derived through the use of compaction vehicles on collection routes, where a 75% reduction of volume is achieved. However, in low-income countries initial compaction are not favourable due to a high initial density of waste. Consequently, compaction vehicles offer little or no advantage and are not cost-effective. Also significant changes in density occur spontaneously as the waste moves from source to disposal, due to scavenging, handling, wetting and drying by weather etc.

Particle Size and Distribution

Since recovery of waste materials is a key element in solid waste management it is important to have knowledge on the size and distribution of the waste constituents. This knowledge is useful in the utilization of mechanical separators and shredders for waste stream processing. However, the particle size and shape of MSW are challenging to measure due to reasons such as their complex shape, difficulty in the movement of MSW particles along the sieve surface and variation in their area depending on the forces acting on it.

The major means of controlling particle size is through shredding. Shredding increases homogeneity, increases the surface area/volume ratio and reduces the potential for preferential liquid flow paths through the waste. Particle size will also influence waste packing densities, and particle size reduction (by shredding) could increase biogas production through the increased surface area available to degradation by bacteria. However, the flip side is, the smaller particles allow higher packing density which decrease water movement, bacterial movement and the bacterial access to substrate. Therefore, it is important for the particle size to be in line with the treatment method to be adopted.

Field Capacity

The field capacity of solid waste is the amount of moisture that can be retained in a waste sample subject to the downward pull of gravity. The field capacity of waste materials is of critical importance in determining the formation of leachate in landfills. Water in excess of the field capacity will be released as leachate. The field capacity varies with the degree of applied pressure and the state of decomposition of the waste. The field capacity of uncompacted commingled wastes from residential and commercial sources is in the range of 50 to 60 percent.

Permeability

Permeability is defined as the hydraulic conductivity of compacted waste. It is an important physical property and it governs movement of liquid and gases in landfill. It depends on pore size, surface area and pore size distribution. Permeability is inversely related to density, implying that denser refuse is less permeable. The reported range of permeability of refuse is 10^{-1} to 10^{-5} cm/sec.

Chemical Characteristic of MSW

Chemical composition of solid wastes is important while evaluating alternative processing and recovery options. Especially while looking at waste to energy processes where waste is used as fuel it is important to have knowledge of proximate and ultimate analysis of the substrate.

Proximate Analysis

A typical proximate analysis includes moisture, ash, volatile matter, and fixed carbon contents.

(a) Moisture: The moisture content of the sample is determined by drying 1 gram of sample at 105° C for one hour. Weight loss is expressed in % of initial weight of sample as:

Moisture (%) = (weight loss/weight of sample) ×100.

(b) Volatile matter: Volatile matter is the weight loss obtained on heating 1 gm sample of substance at 950 °C for 7 minutes in the absence of air:

Weight loss due to VM = Total weight loss – moisture.

Volatile Matter (%) = (weight loss due to VM/ weight of sample) ×100.

(c) Ash: Ash is the weight of the residue obtained after complete combustion of one gram of the substance at 700-750 °C.

Ash (%) = (weight of residue/weight of sample) × 100.

(d) Fixed carbon: Fixed carbon is the material, other than ash, that does not vaporize when heated in the absence of air. It is usually determined by subtracting the sum of the first three values (i.e) moisture, ash, and volatile matter in weight percent from 100 percent.

Fixed carbon (W% wet basis) =100– (%M+%Ash + %VM).

Table summarizes the proximate analysis of typical municipal solid waste generated. The moisture content of municipal solid waste is in the range of 15-40%. Likewise, the volatile matter and fixed carbon is 40-60% and 5-12%, respectively.

Table: Proximate analysis of municipal solid waste.

S. No	Components	Value percent	
		Range	Typical
1	Moisture	15-40	20
2	Volatile matter	40-60	53
3	Fixed carbon	5-12	7
4	Glass, ash and metal	15-30	20

Fusing Point of Ash

The fusing point ash is defined as that temperature at which the ash resulting from the burning of waste will form a solid (clinker) by fusion and agglomeration. Typical fusing temperature for the formation of clinker from solid waste range from 1100 to 1200 °C.

Ultimate Analysis of Solid Waste Components

The ultimate analysis of a waste component typically involves the determination of the percent C (carbon), H (hydrogen), O (oxygen), N (nitrogen), S (sulphur), and ash. Due to the concern over the emission of chlorinated compounds during combustion, the determination of halogens is often included in an ultimate analysis. The results of the ultimate analysis are used to characterise the chemical composition of the organic matter in MSW. They are also used to define the proper mix of waste materials to achieve suitable C/N ratios for biological conversion processes. Analysis for solid waste for carbon, hydrogen, nitrogen and sulphur can be done using CHNS analyser.

Heat Content

Calorific value is the amount of heat generated from combustion of a unit weight of a substance, expressed as kcal/kg. The calorific value is determined experimentally using Bomb calorimeter in which the heat generated at a constant temperature of 250 °C from combustion of a dry sample is measured. Since the test temperature is below the boiling point of water, the combustion water remains in the liquid state. However, during combustion gases remain above 1000 °C so that the water resulting from combustion is in the vapour state.

Table: Heating value of municipal solid waste.

Components	Heating value	
	Range	Typical
Food waste	3500-7000	4500
Paper	11500-18500	16500
Cardboard	14000-12	16000
Plastic	28000-37000	32500
Textiles	15000-20000	17500
Rubber	21000-28000	18500
Leather	15000-20000	17500
Garden trimmings	2300-18500	6500
Wood	17500-20000	18500
Glass	120-240	140
Non ferrous metal	240-1200	700
Ferrous metal	240-1200	700
Dirt, ash and bricks	2300-11500	7000
Municipal solid waste	9500-13000	10500

Essential Nutrients and Other Elements

Where the organic fraction of MSW is to be used as feedstock for the production of biological conversion products such as compost, methane, and ethanol, information on the essential nutrients and elements in the waste materials is of importance with respect to the microbial nutrient balance and in assessing what final uses can be made of the materials remaining after biological conversion.

Biological Characteristic of MSW

The most important biological characteristic of the organic fraction of MSW is that almost all of the organic components can be converted biologically to gases and relatively inert organic and inorganic solids. The production of odours and the generation of flies are also related to the putrescible nature of the organic materials found in MSW (e.g., food wastes). The organic fraction of MSW (excluding plastic, rubber and leather) includes water-soluble constituents (such as sugars, starches, amino acids, and various organic acids), hemicellulose, cellulose, Fats, oils, and waxes, lignin, lignocellulose and proteins. While evaluating incineration as a means of disposal or energy recovery, it is important to keep in view the following facts:

- Organic material yields energy only when dry.
- The moisture contained as free water in the waste reduces the dry organic material per kilogram of waste and requires a significant amount of energy for evaporation.
- The ash content of the waste reduces the proportion of dry organic material per kilogram of waste. It also retains some heat when removed from the furnace.

Waste Processing

Waste processing includes a lot of different processes. Processing, by definition, falls under the heading of physical treatment by using physical techniques for changing the composition and character of the waste. It is possible to distinguish between thermal and mechanical treatment. Every recycling process represents a conversion of the material. So the process of recycling is also a process of production of new materials. It describes the process of producing secondary raw material and compares a series of activities that involves collecting, sorting, processing or converting used materials into useful goods. All these activities conduce to reduce the amount of waste which has to landfill. So the goal of every recycling process is to use or reuse materials from garbage in order to minimize the amount of waste. Of significance in this recycling process is that the waste processing starts with the separation. A separation of waste at the beginning of the process supports the treatment of the waste and helps to raise the efficiency of the facilities which deals with the waste.

Size Reduction

Size reduction is the process in which the particle size of a substance is reduced to a finer state (smaller particles to crude particles or to fine particles) of desired shape and size by using external forces. Size reduction process is also known as comminution and grinding.

Classification of Size Reduction

On the basis of feed and size of the particles, size reduction is classified into 3 types namely, coarse size, intermediate size and fine size reduction. They are as follows:

- Coarse Size Reduction: Coarse size reduction is mainly employed for reducing the size of rigid and unevenly formed solid substances to about 2 – 96 inches or higher. The equipments used for coarse size reduction are jaw crusher, gyratory crusher, toothed roll crusher, hammer mill.
- Intermediate Size Reduction: Intermediate size reduction is mainly employed for reducing the size of rigid and unevenly formed solid substances to about 1 – 3 inches. The equipments used for intermediate size reduction are cone crusher, crushing rolls, stamp hill, disintegrator.
- Fine Size Reduction: Fine size reduction is mainly employed for reducing the size of rigid and unevenly formed solid substances to about 0.25 – 0.5 inches. The equipments used for fine size reduction are ball mill, tube mill and rollers mill.

Factors Affecting Size Reduction

The various factors that affect size reduction are hardness, robustness, abrasiveness, glueyness, softening temperature, particle nature, moisture level, purity requirement, ratio of feed size to product ratio, bulk density and physical effect.

Hardness

The surface property of a material is hardness. Moh's scale is considered as a random scale of hardness. The hardness number ranges between 1 and 10 for a number of mineral substances (for instance from graphite to diamond). Materials with hardness number ranging between 1-3 are called as soft and could be spotted with a fingernail. Materials with hardness number beyond 7 are hard and could not be spotted with a blade. The materials having hardness number ranging between 4-6 are designated as intermediate. Normally, when the material is harder, then it is complicated to reduce the size of the material.

Robustness

Robustness of a material is also an important factor that affects size reduction. In some instance, it is considered as more important than hardness of a material. A soft and tough material may lead to many problems in size reduction when compared to hard and delicate material. For instance, it is harder to fracture rubber than a blackboard chalk. Toughness creates difficulty in many materials, particularly in fibrous materials and is often connected with moisture content of a material.

Abrasiveness

Abrasiveness is an important property of hard materials (principally those of mineral origin) and could limit the efficiency of equipment that could be employed. During the milling of some very rough materials, the final fine particles may be polluted with in excess of 0.1 % of metal tattered from the pulverizing grinder.

Glueyness

Stickiness of a material is another property that leads to substantial complexity in size reduction of a material. The sticky substance may stick to the milling plane, or on the lattice of the screen could turn to be choked. If heat is produced during any size reduction process, then the sticky and resin like materials may cause problems to the entire process. By adding inert substances it can provide assistance for such process (For example, kaolin to sulphur).

Softening Temperature

In size reduction process, heat is produced occasionally. The produced heat may lead to softening of some substances. The temperature at which the heat generating process happens is very essential. Waxy substances, such as stearic acid, and oils or fats containing substances are some of the substances that may be influenced by this temperature. To overcome these drawbacks, some process can be used to surmount this effect such as cooling of milling equipment either by using water jacket or allowing a stream of air to pass through the equipment. Another option is to employ liquid nitrogen.

Particles Nature

Some of the materials are homogenous in nature. However, many materials possess various unique structure, for instance, mineral substances found to be weak in nature and these substances crack to produce flake-like elements, while other substances such as vegetable wastes being cellular nature have fibrous like structure.

Moisture Level

Moisture level is considered as an essential parameter that posses some major characteristics which can affect size reduction, for instance, hard nature, robustness or glueyness. Normally substances must be dried out or damp. Substances that possess less than 5 percent of moisture content are considered to be appropriate for dry grinding process. Greater than 5 percent of moisture content leads to agglomeration or pasting. In the same way, substances that have more than 50 percent of moisture content are considered to be appropriate for wet grinding process. In case of low moisture content materials, coarse and intermediate size reduction process is suitable. In case of high moisture content particle, greater than 50 percent, fine size reduction process is applicable.

Purity Requirement

Some of the size reduction equipments make the crushing or milling surfaces to worn and those processes must be avoided in case of high purified product requirement. Likewise, some equipment is considered to be inappropriate if the cleaning of various materials is complicated.

Ratio of Feed Size to Product Ratio

It is the ratio of feed to particle size (diameter) of product. Equipments that are used for producing very small sized particles could be essential to perform the size reduction process in various phases with the help of diverse machines, for instance, primary process such as crushing. Subsequently the crushed materials are subjected to coarse grinding process and finally subjected to fine grinding. For coarse size reduction process, the reduction ratio should be between 3 to 7. For grinding mill, the reduction ratio should be greater than 100.

Bulk Density

The potency of many batch mill processes relies on volume. These processes normally require solid substances on the basis of weight. Therefore, all the parameters are kept equal and the substances that are produced from the equipments after the completion of process is connected with the bulk density of the materials.

Physical Effect

Some of the materials are very effective and powerful. Minute quantities of dust particles can influence the operators. To eliminate these drawbacks, milling equipments must be covered. Systems that can extort air are attractive and beneficial. Wet grinding is appropriate as it eliminates this issue completely.

Energy and Force Applied for Size Reduction Process

In size reduction process, the increment in surface area is enhanced by the applied energy. The input energy is utilized to cause the fracture and defects in the materials. The remaining energy is utilized for elastic twisting, transportation of materials for generating friction forces between the grind mill and material and fractioning forces and noise. About 0.1-2% of the energy supplied to the equipments is arisen as the increased energy of the subjected material. Surface energy estimates the extent of intermolecular bonding destruction of a material during the formation of new surface. This can be estimated through a number of methods by contact angle goniometer. Effectiveness of size reduction process depends on the load application step and its extent. Extent and nature of force applied is also an important factor to be taken into account during size reduction process. For example, when the force input is inadequate for the elastic limit to be prevailed and the particles are compacted, then the energy will be piled up in the particle.

Once the applied load is removed, then the particle subjected to size reduction process come to its original condition. In that case, the energy is generated as heat. Application of greater force will cause the particles to get crack. To utilize the energy adequately, the force applied should be slightly greater than the grinding potency of the substances. During size reduction process, the surface of the particles is generally uneven in nature and therefore the force is mainly focused on the elevated spots. As a result, greater pressure and elevated temperature is experienced by the material closely. When the particle is cracked slightly then the spot at which the force applied is get altered. Additionally, the kinetic energy is related with the rapid acceleration of material experiences a crack. The degree of surface fracture energy extent per unit volume of various materials are glass 1–10 J/m2, plastics 10–103 J/m2 and metals 103–105 J/m2. All these surface fracture energy values are found to be greater than the degree of thermodynamic surface energy extent which is approximately 10-1 J/m2.

Mechanism of Size Reduction

During size reduction process, smaller particles are produced from larger particles of the same material. Size reduction leads to the formation of crack in the larger particles. The basic three mechanisms involved in size reduction are cracking, crushing and grinding of solid particles. The particles are fractured through external shear forces (mechanical energy). Then the particles are crushed into smaller coarse particles. Finally, the crushed particles are grinded through grinders. During this process, by the application of mechanical energy the particles are distorted, loosened and cracked.

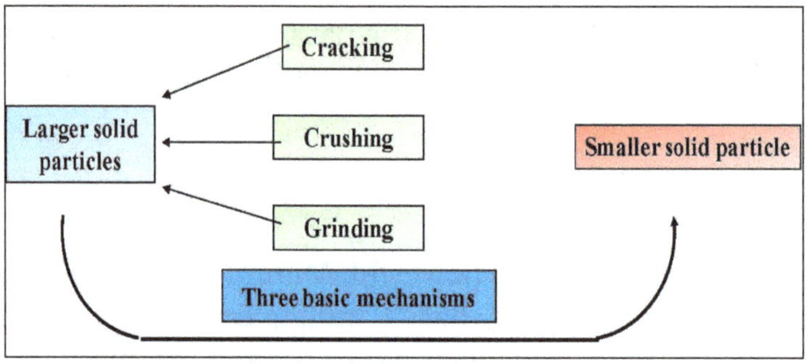

Figure: Basic mechanism of size reduction.

There are various mechanisms involved in size reduction process. They are impact, compression, shear and attrition:

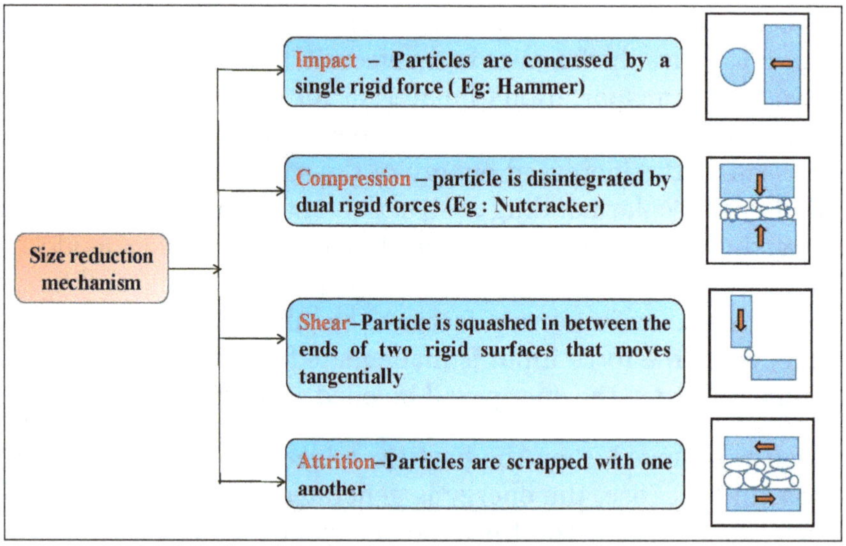

Figure: Different mechanism of size reduction.

- Impact: Impact is a mechanism which occurs when a solid particle (almost immobile) is beaten by a highly accelerated moving substance or when a moving substance hits an immobile surface.
- Compression: Compression is a mechanism which occurs when a solid particle is disrupted by two rigid surfaces.
- Shear: Shear is a mechanism which occurs when a solid particle is compressed between rollers through mechanical pressure.
- Attrition: Attrition is a mechanism which occurs when the materials are subjected to force similar to shear but the surfaces scrapped with one another. As a result, shear forces are generated which shatter the materials.

Size Reduction Laws

The energy requisite of size reduction process mainly depends upon the parameters such as input and output of particle size, rigidity, force and other properties of solids. The different laws that describes energy need of particle size reduction process are:

- Kicks law,
- Rittinger law,
- Bond's law.

Kicks Law

Energy needed to crush a particular material is invariable for the similar reduction ratio irrespective of its original size (Energy is directly proportional to ratio of size change).

$$Hp = K \log \frac{D}{d}$$

Hp = Energy, K= constant, D= initial diameter, d = final diameter. Energy required to crush material of 1 inch into 0.5 inch = Energy required to crush material of 3 inch into 1.5 inch. Application: Kick's theory is suitable for compression of larger materials.

Rittinger Law

Energy needed to crush a particular material is proportional to the produced new surface area. (Energy is directly proportional to the new surface area produced).

$$Hp=K\left(\frac{1}{d}-\frac{1}{D}\right)$$

Application: Rittinger's theory is suitable for delicate materials which undergoes fine grinding.

Bond Law

Energy needed to crush a particular material into a desired particle size is proportional to the square root of product's surface area to volume ratio. (Energy utilized in fracture propagation is directly proportional to produced fracture extent),

$$Hp=100E_i\left(\frac{1}{\sqrt{d}}-\frac{1}{\sqrt{D}}\right)$$

E_i = Bond's working index.

Application: Bond theory is applicable to particles which undergoes rough grind sizing. In bond theory, the working index is used for evaluating the effect of grinding functions.

Size Reduction Equipments

The machines used for reducing the particle size of materials are called as size reduction equipments. They are of various types. The most common types of size reduction equipments are shredders, glass crushers and wood grinders. Table summarizes the different equipment used for size reduction.

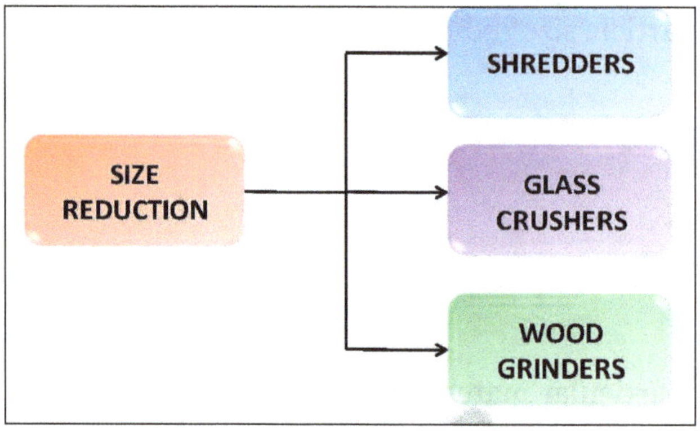

Figure: Various size reduction equipments.

Table: Size reduction equipment.

Type	Mode of action	Application
Slight grinders	Milling, crumbling	Organic possessed solid wastes.
Chippers	Cutting, slicing	Paper, cardboard, tree cuttings, yard waste, wood, plastics.
Bulky grinders	Milling, squeezing	Breakable and crumble materials, used mostly in industrial purpose.
Jaw crushers	Mashing, crumbling	Huge solids
Rasp mills	Milling, tearing	Disposed solid wastes.
Shredders	Cutting, tearing	All types of municipal wastes.
Cutters, Clippers	Cutting, tearing	All types of municipal wastes.
Hammer mills	Crumbling, cutting, slicing	All types of municipal wastes, most generally used equipment for size reduction and adjusting constitution of wastes.
Hydropulper	Cutting, tearing	Ideally suited for use with pulpable wastes, including paper, wood. Used initially in the paper making industry. Also used to destroy paper accounts.

Shredders

Shredders are the equipments used to shred the feed materials with the help of rotating shaft or sharp knives. These equipments are mainly used to reduce the size of municipal solid wastes. Examples of shredding equipments include: hammer mill, flail mill, shear shredder, cutting mills, cage disintegrators, roller mills and hydro pulpers. The hammer mill is the most commonly used shredding device.

Hammer Mill

Hammer mill is a type of shredding device that works under the principle of collision between the fastly moving hammer located on a rotor and the solid material. During operation, the hammers of hammer mill which is fitted to a rotary component, hit the solid materials that is introduced into the device and shred it. In hammer mills, the rotor rotates at a speed of 8000 -15000 rpm. Once the waste material is shredded, they are pulled via the exit of the device. The exit section (discharge unit) may or may not consist of bottom grates. The hammer mills are of two types namely, One-way type and reversible type. In one-way type hammer mills, the breaker plates are replaceable. In reversible type, the breaker plates are reversible (adjustable).

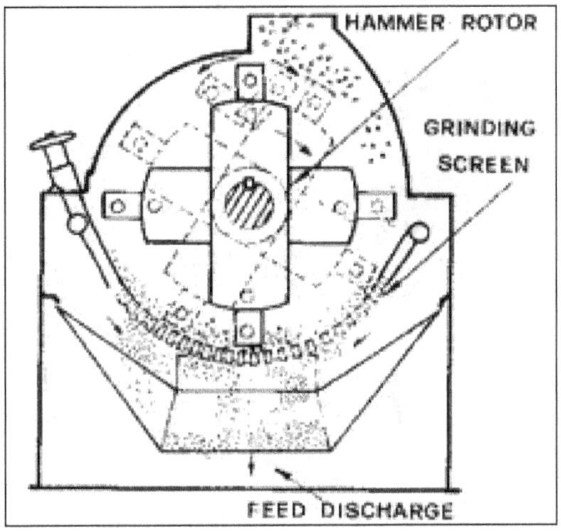

Flail Mill

The flail mills are solitary pass shredding equipment works similar to the hammer mill, however afford only coarse tearing of materials, since the hammers are spaced further apart. During operation, the waste materials stay in the hammer till it moves via the hole in the bottom grate. They are often utilized as bag breakers.

Shear Shredders

Shear shredders are the equipments in which the shredding of waste materials is take place by the application of shear forces. The shear shredder consists of two analogous counter rotary shafts with a sequence of discs located vertically which serve as shears. The solid particle to be shattered is moved to the middle of the counter rotary shafts. The magnitude of the solid particle is reduced by the shear action of the shattering discs. The shredded particles is forced via the outer exit section. Shear shredders are also used as bag breakers.

Hydropulpers

Hydropulper is also a type of shredding device mainly used for reducing the size of municipal solid wastes. During operation, the municipal solid waste material and water to be shredded are moved to the device. The fastly revolving shaft part slices and transfers the solid content into pulpable finish products. The resulting pulpable end products consist of 2.5 to 3.5 percent of solid particles. The non-pulpable substances are collected from the side tank of hydropulper unit. The remaining slurry is discharged through the base of the unit with the help of a pump and subjected to further processes. The waste effluent substances are settled down in the base of unit by gravity with bucket elevator.

Glass Crushers

Glass crushers are the devices employed to mash glass vessels and other glass materials present in municipal solid wastes. The glass crushers are works under the principle of compression by the application of stress forces. The stress forces are applied by rotator wheels or rollers. The glass crushers consist of two rollers, front and back roller. A compression spring in the unit provides stress forces.

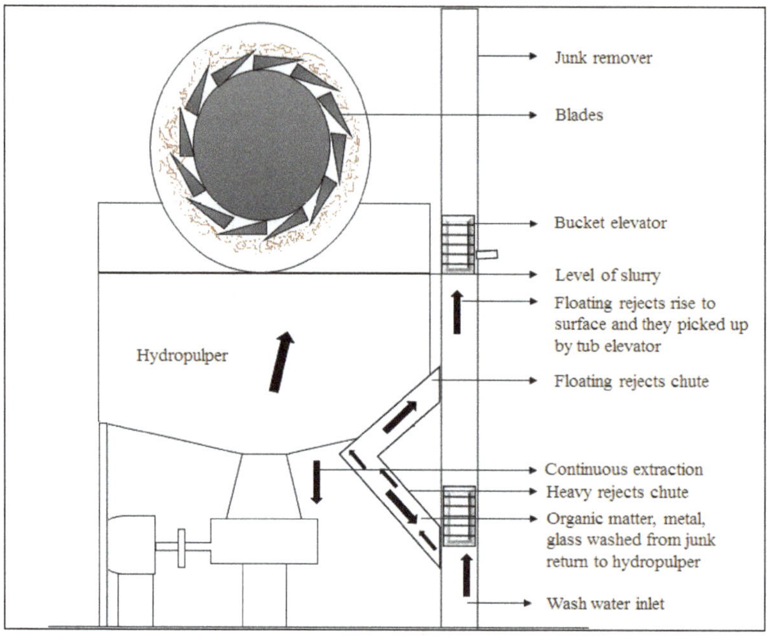

During operation, the rollers are allowed to rotate. The materials to be compressed are introduced through the gap between the rollers. Through the application of high pressure, the materials are trampled. Glass is always compressed once it is segregated to minimize the cost incurred towards storage and shipping. In some mechanical segregation process, glass is trampled, subsequent to several segregation phases, to influence its elimination by screening. Crushed glass can be removed optically based on colour.

Wood Grinders

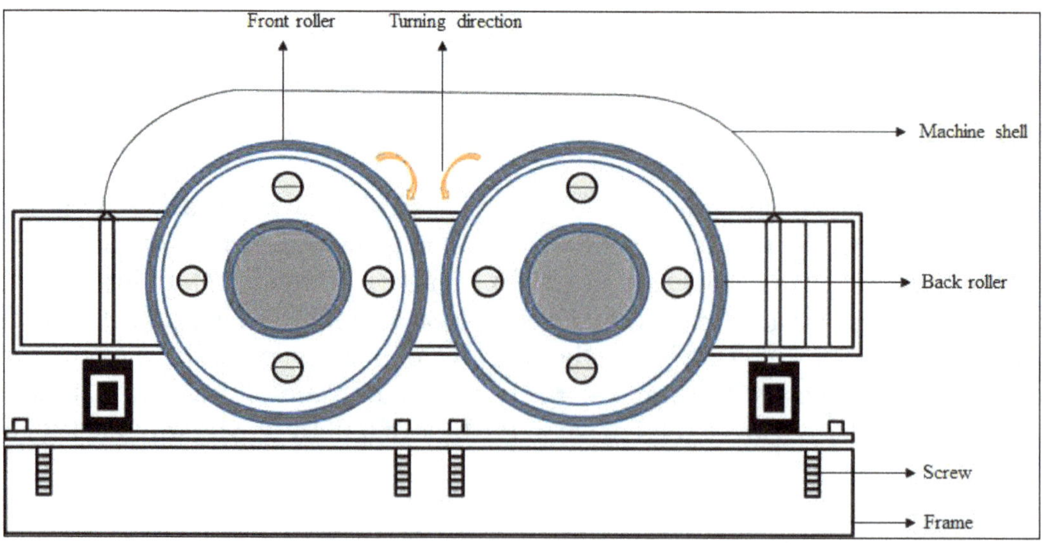

In general, most of the wood grinders are wood chippers and are used to grate large wooden logs into chips. This could be utilized as a fuel and fine substances. The finer material in turn can be composted easily.

Wood grinders consist of a rotating apex part (rotating arm), and a immobile base part holding a hammer mill. The wooden pieces to be grinded are fed through a hopper and the rotating force of the rotating arm makes certain that the substances run incessantly to the hammer mill. The incessant flow of grinded substances is pulled speedily from the grinder through a conveyor at the bottom.

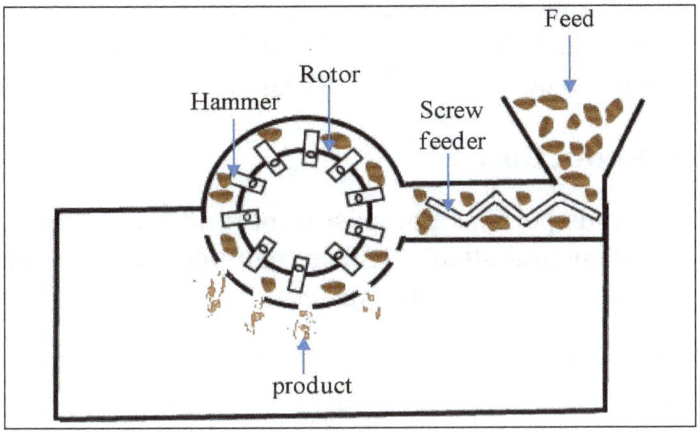

Figure: Wood crusher.

Selection of Size Reduction Equipment

The following aspects should be taken into account while selecting size reduction equipment: The nature and properties of materials should be considered. Size requirements for tattered particles should be considered. The process involved in feeding section, condition of suitable milling ability should be considered. The operation condition (whether it is continuous or intermittent) should be taken into account. Operational characteristics such as energy requirement, permanent finely equipped maintenance constraints, ease mode of action, consistency, noise creation, air and water decontamination should be properly maintained. Land accessibility, height, noise and environmental restrictions should be considered. Storage of ferrous materials for additional processes should be considered.

Safety Issues

The equipments employed for tear up action contain volatile substances like solvents and gasoline. The volatile substances will generate highly dangerous exhaust. The hammer mill shredder could produce sparks, when operating with high revolution on metals. It leads to a serious explosive issue. These are the major issues related with size reduction process.

Safety Systems

Size reduction equipment (Shredder) should be placed in a segregated way when compared to other operating sections. Volatile proof substances must be used for electrical switches, mains, wiring and lightings to circumvent short circuiting. The room that contains shredder device must have an explosion ventilator to protect the building. An automatic fire controlling device should be installed to minimize the explosion level at initial stage of operating the devices. During explosion, the inert gases such as nitrogen or carbondioxide must be sprayed for a fraction of seconds to avoid fire problems. Halogens can also be used as the fire extinguishing agent.

Volume Reduction

Volume reduction is a type of waste processing where the nature of the waste is altered physically. It is one of the essential steps in waste management system. Proper waste processing will improve the efficiency of solid waste management; facilitate recovery of resources and energy conversion. Volume reduction is defined as handling or processing of waste materials to reduce the volume of the waste. Volume reduction can be achieved by three means:

Mechanical Volume Reduction

Mechanical volume reduction is perhaps the most important factor in the development and operation of solid waste management systems. Vehicles equipped with compaction mechanisms are used for the collection of most municipal solid wastes.

To increase the useful life of landfills, wastes are compacted. When compacting a broad range of municipal solid wastes, it has been found that the final density is essentially the same regardless of the starting density and applied pressure. This fact is important in evaluating the claims made by manufacturers of compacting equipment.

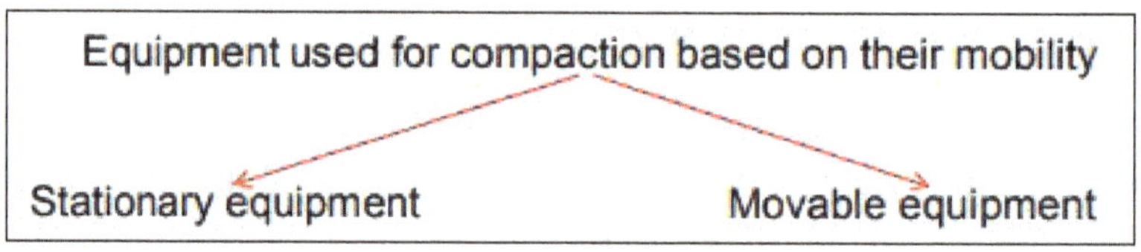

Stationary Equipment

Equipment in which wastes are brought to and loaded into either manually or mechanically. According to their application, stationary compactors can be described as light duty, commercial or light industrial, heavy industrial, and transfer station compactors. Large stationary compactors are necessary when wastes are to be compressed into steel containers, or transport vehicles directly.

Movable Equipment

This represents the wheeled and tracked equipment used to place and compact solid wastes as in a sanitary landfill.

Shredding

This is required to convert large-sized wastes (as they are collected) into smaller pieces. Size reduction helps in obtaining the final product in a reasonably uniform and considerably reduced size in comparison to the original form.

It is noted that Size reduction does not necessarily imply volume reduction and this must be factored into the design and operation of SWM systems as well as in the recovery of materials for reuse and conversion to energy. In the overall process of waste treatment and disposal, size reduction is implemented ahead of:

- Land filling to provide a more homogeneous product. This may require less cover material and less frequent covering than that without shredding. This can be of economic importance, where cover material is scarce or needs to be brought to the landfill site from some distance.
- Recovering materials from the waste stream for recycling.
- Baling the wastes: a process sometimes used ahead of long distance transport of solid wastes – to achieve a greater density.
- Making the waste a better fuel for incineration waste energy recovery facilities. (The size reduction techniques, coupled with separation techniques such as screening, result in a more homogeneous mixture of relatively uniform size, moisture content and heating value and thereby improving the steps of incineration and energy recovery).
- Reducing moisture, i.e., drying and dewatering of wastes.

Type	Mode of Action	Application
Small grinders	Grinding, mashing	Organic residential solid wastes.
Chippers	Cutting, slicing	Paper, cardboard, tree trimmings, yard waste, wood, plastics.
Large grinders	Grinding, mashing	Brittle and friable materials, used mostly in industrial operation.
Jaw crushers	Crushing, breaking	Large solids.
Rasp mills	Shrdding, tearing	Moistened solid wastes.
Shredders	Shearing, tearing	All types of municipal wastes.
Cutters, clippers	Shearing, tearing	All types of municipal wastes.
Hammer mills	Breaking, tearing, cutting, crushing	All types of municipal waste, most commonly used equipment for reducing size and homogenizing composition of wastes.
Hydropulper	Shearing, tearing	Ideally suited for use with pulpable wastes, including paper, wood chips. Used primarily in the papermaking industry.

Chemical Volume Reduction

Chemical volume reduction is a method, wherein volume reduction occurs through chemical changes brought within the waste either through addition of chemicals or changes in temperature.

Incineration is the most common method used to reduce the volume of waste chemically and is used both for volume reduction and power production. These other chemical methods used to reduce volume of waste chemically include pyrolysis, hydrolysis, and chemical conversions.

Thermal Volume Reduction

The volume of municipal wastes can be reduced by more than 90% by incineration. In the past, incineration was quite common. However with more restrictive air pollution control requirements necessitating the use of expensive cleanup equipment only a limited number of municipal incinerators are currently in operation.

More recently, increased haul distances to available landfill sites and increased fuel cost has brought about a renewed interest in incineration and a number of new incinerator projects are now on the drawing boards.

Compaction

Waste compaction is the process of compacting waste to reduce its volume and densify the waste. In simple terms it is the process of pressing the waste to compact and increase the density. Compaction is done at three levels, namely in collection vehicles, transfer stations and disposal site. Garbage compactors and waste collection vehicles compress waste so that more waste can be loaded within the provided space. At the transfer station they are compacted into bales or cubes or pellets. Waste is also compacted again, more thoroughly, at the landfill to conserve valuable airspace and to extend the landfill's life span.

Benefits of Compaction

The benefits of volume reduction by compaction:

- Reduction in the amount of materials to be handled at the disposal site.
- Improved efficiency of collection and disposal of wastes.
- Feasibility analysis of waste management system.

Demerits of Compaction

- Recyclable material quality will be poor, when solid wastes get compacted.
- Once solid waste compacted, the separation or sorting process is too difficult.
- Food wastes and organic bio – degradable matters when compacted along with other waste will reduce the recycling value of paper and plastics.

Equipment used for Compaction

Based on the Movement

The compactors are categorized into two types based on their movement:

- Stationary equipment: Stationary equipment is operated either manually or mechanically to collect the solid wastes and load them into the containers. The collection vehicle equipped with compaction mechanisms for compacting the solid wastes is called stationary compactors. Such compactors can be placed in residential, industrial and commercial areas. They are also placed in transfer stations. Stationary compactors can be classified according to their application as (1) light duty - used for residential and commercial solid wastes; (2) commercial or small industries; (3) large industries; (4) transfer station.

- Movable equipment: This consists of wheel attached vehicle and tracked equipment used to dump the compacted solid wastes in a sanitary landfill.

Based on the Pressure

The compactors used at transfer stations are classified into 2 major categories based on the pressure used for compaction. They are low pressure (less than 7 kg/cm^2 or 100 lb/in^2); ii) high pressure (more than 7 kg/cm^2 or 100 lb/in^2).

- Low pressure (less than 7kg/cm^2) compaction: Large sized container is required for low – pressure solid waste compaction. These low pressure compaction systems are used in many industries in association with material recovery such as paper and cardboards.
- High pressure (more than 7kg/cm^2) compaction: In high – pressure compaction systems, specialized compactors systems are used to minimize the volume of solid wastes into small sections, cubes or bales of various sizes. Pulverized or shredded waste can be compacted to form logs or bales. The nature of waste determines the use of high – pressure compaction equipment. High pressure compaction units possess a capacity to compact waste upto 351.5 kg/cm^2 or 5000 lb/in^2 and the reduction range may vary from about 3 to 1 through 8 to 1.

Bailing Equipment

Bales are another substitute to compaction equipment. It works under huge pressure, mostly 100 to 200 lb/in^2. Through this equipment relatively low, compact bales of solid waste or recycled materials are generated. Bale size ranges from 48 x 30 x 42 in. up to 72 x 30 x 44 in. Likewise, the weight of the bales varies based on the solid waste composition and ranges between 1150 to 1800 lb for the light and high bales of crumpled cardboard. Moreover, baled solid wastes should be landfilled in "balefills". Practically all kind of recycled materials like cardboard, paper, plastic, PETE bottles and aluminium cans be baled and sent to recycling industries. These kind of baled materials are easily load with the help of forklifts and can be economically shipped due to its large bulk density. Balers can be of two types: vertical and horizontal. Vertical balers are used in small scale operations and horizontal balers are used for large scale operations.

Cubing and Pelleting Equipment

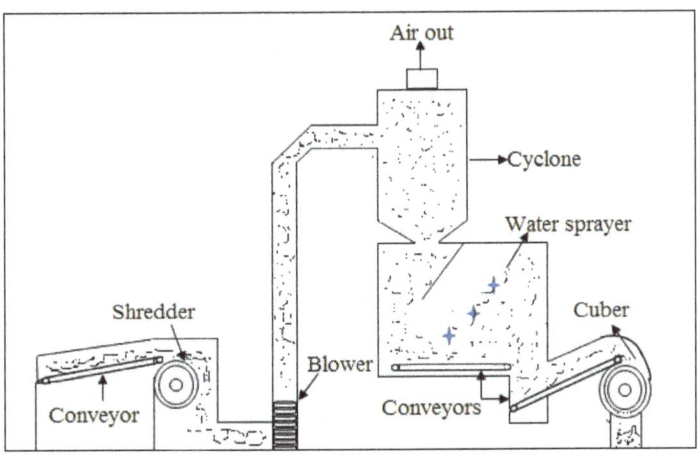

Figure: Cubing (densification system).

Cubing and pelleting is a process that can be used to generate densified refuse derived fuels (dRDF) for combustion in incineration, gasification or pyrolysis techniques. The expenditure of the process can vary according to the following features namely when the dRDF has to be exported to a long distance, stocked or stored for more periods, or fired in existing boilers made for coal or wood. Cubing and pelleting equipment were operated under same principle. Waste paper RDF is ejecting through extrusion dies with an eccentric revolving press wheel. An entire cubing or pelleting techniques demands a shredder, a conveyor and a moisture control system. The cubes or pellets are joined by heat due to friction as the cubes or pellets are ejects. The cubes and pellets stored in dry condition, dRDF can be kept for months without decay. Normal cubes are almost 3 in by 1 in 2. Pellets are circular, basically 0.5 to 0.75 in diameter by 0.5 to 1 in long.

Table: Compaction equipment used at different locations.

Location or Operation	Type of Compactor Stationary/residential	Remarks
Solid waste generation points	Vertical	• The equipment provides a perpendicular compaction using a ram in perpendicular direction. • It may be operated automatically or hydraulically. • They are usually hand-fed. • Wastes compacted into corrugated box vessels or paper or plastic bags. • Popularly used in medium and high-rise flats.
	Rotary	• The Ram device used for densifying the waste is fixed on a revolving platform. • Hammer mechanism used to thicken waste. • The platform rotates as containers are filled with waste. • They used to densify paper or plastic bags. • Used in medium and high-rise apartments.
	Bag or extruder	• Compactor can be chute fed. • Both perpendicular and horizontal rams are used. • Densification is done for both single and multi-bags. • Used in medium and high-rise flats. • Figure depicts bag or extruder type compactors. The revolving press at the center compact the solid waste and forces it through the extrusion dies. Through the opening the waste comes out as cubes.
	Under counter	• They are small compactors. • Used in individual houses and apartment units. • Wastes is dropped into bags through a panel door, then it is compacted into paper bags. • The odour released is controlled by odour controller sprays.
	Stationary/commercial	• The compaction takes place in steel vessels fixed with ram. • The ram can be vertical and horizontal. • The compressed wastes are physically tied and removed. • Used in low, medium and highrise flats, commercial and industrial services.

Collection	Stationary/packers	• Solid gathering vehicles armed with compaction mechanism.
Transfer and/or processing station	Stationary/transfer trailer	• Trailers used for transport are fitted with compactors.
	Stationary low pressure	• Wastes are compacted into large containers.
	Stationary high pressure	• Wastes are compacted into solid bales or other forms.
Disposal site	Movable wheeled or tracted equipment	• Particularly designed to achieve extreme compaction of wastes.
	Stationary/track mounted	• High-pressure transportable motionless compactors used for volume reduction at a discarding site.

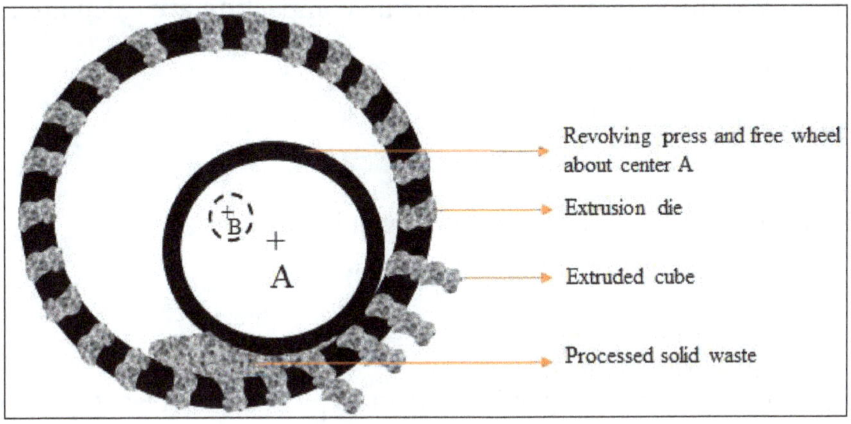

Figure: Bag or extruder.

Criteria for the Selection of Compaction Equipment

To confirm active processing, the following factors should be considered while choosing compaction equipment:

- Characteristics of solid waste: Size, composition, bulk density and moisture content of the solid waste to be compacted.
- Method of transporting, handling and feeding of solid wastes to the compactor equipment.
- Possible usages of compacted solid waste materials.
- Design features such as the size of filling chamber, compaction pressure, compaction ratio, etc.
- Operative characteristics such as energy requirements, maintenance requirement, ease of operation, reliability, noise output, air and water contamination regulator requirement.

Table: Factors governing selection of densification equipment.

Factor	Comment
Scope of densification	Compaction system for processing of solid wastes, bailing for handling recyclables and solid wastes.
Characteristics of material to be processed	Particle size, shape, water content, composition, specific weight.

Transfer technique for materials to and from the densification system	Conveyor characteristics and specifications, forklifts used to handling purpose.
Equipment design parameters	Unit capacity, power demand, compaction ratio, operating pressure.
Operational condition	Energy demands, maintenance, complexity of operation, noise and air emissions.
Site factors	Floor level and vertical space availability, approach.

Compaction Ratio

The compaction ratio and the volume reduced during compaction are calculated as per the equation below:

$$\text{Volume reduction}(\%)=\frac{V_i - V_f}{V_i}\times 100$$

The compaction ration of the waste by equation, given below:

$$\text{Compaction ratio} = \frac{V_i}{V_f}$$

where V_i = volume of waste before compaction, m^3 and V_f = volume of waste after compaction, m^3.

The relationship among the compaction ratio and the percent of volume reduction is depicted in figure. Upto 80% volume reduction, the compaction ratio increase is very slow and gradual. Above which a gradual increase in compaction ratio occurs. To secure an increase from 80 to 90 %, an increase in compaction ratio from 5 to 10 is required. This analysis was important to assess the trade-off analysis among compaction ratio and overall compaction ratio.

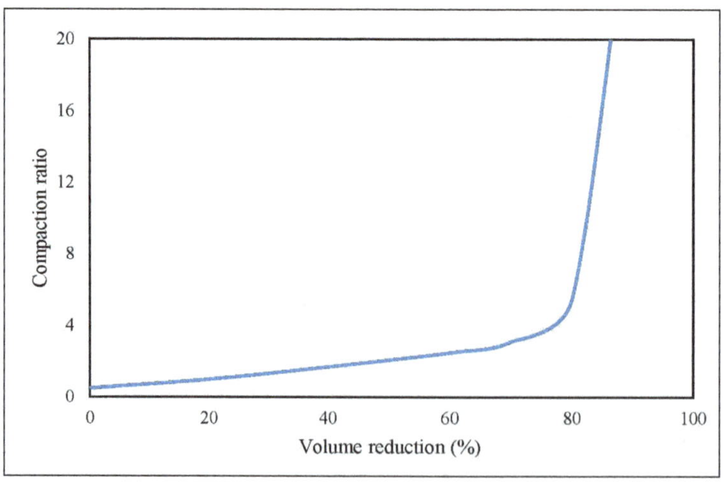

Figure: Compaction ratio vs volume reduction.

Another imperative factor that should be concluded is the final specific weight (density in metric units) of the waste after compaction. Some regular curves for raw solid waste are presented in Figure the asymptotic value used in expanding these curves in 1800 lb/yd³, which is steady with values attained by using high pressure compactors. As a point of reference, the specific weight of water is 1685 lb/yd³.

When shredded wastes are compacted under the same scenario, the density may be up to 35 percent greater than that of raw wastes, up to an induced pressure of 100 lb/in². The higher specific weight obtained under the applied pressure greater than 100 lb/in² is not changed significantly by shredding.

Perhaps the most significant fact to be noted in Figure is that the primary raise in specific weight brought about by the application of pressure are mostly dependent on the initial specific weight of the waste to be compacted. This case is relevant when claims made by producers of compaction machinery are examined. The moisture content, which changes with region, is another variable that has a huge reaction on the degree of compaction obtained. In some stationary compactors, arrangement is made to introduce moisture, basically in the form of water, during the compaction mechanism.

For cubing and pelleting equipment, the two imperative achieved characteristics are the unit and bulk specific weights of the cubes and pellets. The unit specific weight is the specific weight of an individual cube or pellet. The bulk specific weight is an analysis of a permanent volume of cubes or pellets and is suggestive of the specific weight of cubes or pellets in the presence or absence of movement because it incorporate the void space caused by randomly packing cubes in a container. The cubing system has obtained unit specific weights of 1750 lb/yd³ for shredded newsprint. However, the bulk specific weight of the cubes was only 814 lb/yd³.

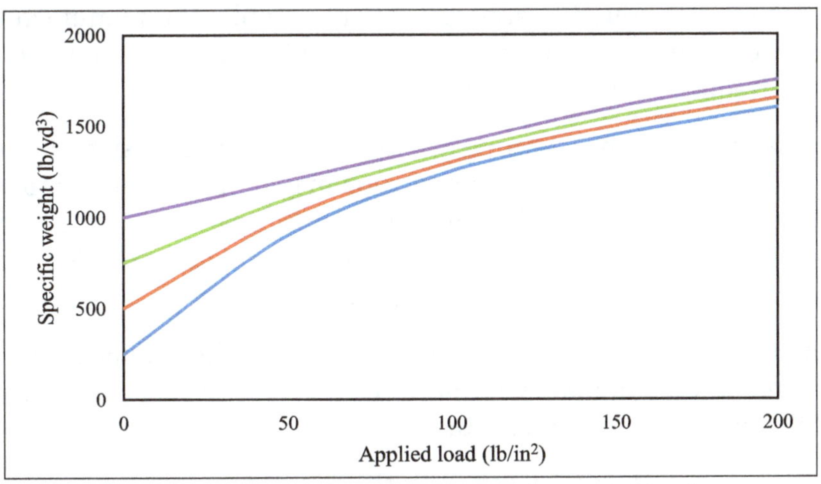

Figure: Specific weight of solid wastes vs applied load.

To make a trade – off analysis among compaction ratio and cost of compacted solids, the relationship between the percent of volume reduction and compaction ratio is very important. Moisture is another important parameter that needs to be considered. The moisture content will vary with respect to location. In certain cases, addition of moisture in the form of water to stationary compactors becomes essential while under processing.

Source Reduction

Source reduction is also known as waste prevention. The purpose of source reduction is to decrease the amount and toxicity of any material that requires management through prevention at the source of generation.

Source Reduction Strategies

Source reduction targets waste at its source, making it more beneficial than recycling or reusing products. By decreasing the amount or toxicity of waste entering the waste stream, we can reduce the impact that waste has on our communities and the environment.

Sometimes opportunities for source reduction are overlooked or unrealized due to existing regulations that focus on treatment and disposal. This is unfortunate because source reduction is more desirable; it nips the issue in the bud instead of continuing to deal with an established problem. For example, a couple of clients have switched from traditional paints to powder coating. This change has had multiple benefits, including improved product performance, dramatically reduced air emissions (reducing their air data reporting), and a significant reduction in hazardous waste. If you're looking to decrease the amount of waste or toxicity produced by your facility, consider the following source reduction strategies:

- Eliminate nonessentials: Sometimes you can eliminate a product, process, or procedure without harming quality or function. To do this, you may need to rethink your needs. Does your company truly need X? Would your customers suffer without Y? Just because you have used X and Y for years, that doesn't mean you can't eliminate them to reduce waste and streamline your production process.
- Improve operating practices: Reassessing an operating practice can allow a manufacturer to implement source reduction strategies. For example, the manufacturer may find that they can reduce the amount of materials needed, eliminate the number of operation cycles, or reduce the frequency of changeovers.
- Modify products: Sometimes a process or product can be redesigned to use fewer materials. For example, by upgrading equipment, you may reduce the loss of material due to contamination, leaks, or spills. Spill kits and drain covers can also reduce spills from escaping in high-risk areas. Or, you may be able to switch from a less efficient material to a more efficient material, reducing the amount of product needed.
- Purchase more durable products: When you invest in a high-quality, durable product that will last for years, you eliminate your need to constantly replace it. This extends the useful lifespan of the product and reduces waste.
- Replace materials to reduce toxicity: Sometimes a less toxic product can be used as a substitute for a more toxic product without reducing quality. For example, a rubber product manufacturer has been adjusting its formulations to replace di (2-ethylhexyl) phthalate with environmentally friendly alternatives.

Effects of Source Reduction

Source reduction possesses both economic and environmental advantages. It reduces the pollution load, minimizes the cost of waste management and disposal. Primarily, source reduction ends up in changing the composition of solid waste.

Economic Benefits

Source reduction reduces the cost of solid waste management through reduction in quantity of waste, avoiding the purchase cost and generating revenue from resale of items. The substantial saving can be achieved when an item is repaired and reused. For example, use of reusable utensils and dishware will minimize or avoid the need of disposal products. Regular maintenance increases the life span of a product thereby minimizing the cost of disposal or replacement.

Sometimes renting or sharing a material during an occasion avoids the purchase and disposal cost of particular item. For example, a computer that become outdated and does not serve the purpose of the owner can be leased to other companies who need them. Material exchange also helps in source reduction. Waste products of one industry can be used as a raw material for another. Example, in a garage the items that are not required are sold out to generate revenue. Many industries save money by reducing packaging material, minimizing waste generation during manufacturing process and using scrap materials in the manufacturing process. Paperless communication example, electronic mail reduces the usage of paper, minimizes paper manufacturing, reduces paper recycling cost and disposal cost.

Environmental Benefits

Source reduction is associated with a number of environmental benefits:

- Reduces the requirement of natural resources.
- Reduces the requirement of energy.
- Minimizes the pollution.
- Reduces the amount of waste sent to waste processing and disposal site.
- Minimizes the greenhouse gas emissions.

Greenhouse gases such as methane, carbon dioxide and oxides of nitrogen are generally released during the mining of raw materials, transport, manufacturing process and disposal of waste residue. For example, during the life span of paper, greenhouse gases are emitted during cutting of trees, combustion of paper and disposal. The level of greenhouse gas emissions from the waste material is very uncertain and hence cannot be quantified. Collection of beverage containers and water bottles and depositing them to the company will reduce litter and collection cost.

At certain times, the environmental benefits of source reduction show negative consequences. For example, using repaired old electronic appliances results in consumption of more energy. Likewise avoiding disposable items might increase the water usage for cleaning plates, cups and other utensils. Further it will also increase the organic and suspended solid load in the waste water. Life cycle assessment can help in determining the actual impact of solid waste management system on the environment.

Waste Composition

The quantity and composition of solid waste is greatly determined by the habits of the community. Switching to lighter packing material or choosing a packing material that can be easily recycled will also result in changes in composition of waste. For example, source reduction in packaging industry: the weight of aluminum cans used in beverage packing is reduced by almost 502% in the last decades. Source reduction can also be done through material substitution. For example, the aluminum cans can be replaced with PET bottles.

Source Reduction Planning

To formulate an effective source reduction program, planning is very essential. A planner should have an idea about the following:

- What needs to be reduced.
- How much reduction needs to be achieved.
- The methods of measuring the results of source reduction.

An effective planning can be done under independent leadership, authority, adequate staffing and appropriate budget. An infrastructure is essential for supporting the source reduction program. The first step in planning includes formulation of a policy with clear definition of terms, source reduction means and its differentiation from recycling and other waste management options. The second step would be to set a goal and establish measurement strategies. These may help in evaluating the progress of the source reduction plan. Four steps are used to set up goals and establish measurement strategy:

- Set up a source reduction goal with specification: Details of baseline year, target year, nature of reduction should be specified.
- Formulate separate goals for the various generating sectors like, residential, commercial, industrial, institutional etc.; materials such as paper, glass, plastic, organics; products such as cardboard boxes, newspapers, glass bottles, cups etc.
- Selection of measurement unit: The measurement unit that is to be used for source reduction program such as weight, volume, weight and volume are decided.
- Selection of suitable measurement methodology: The success of source reduction is measured through waste audits, service, sales tracking and sampling.

Information Required for Measuring Source Reduction

Data on population, projection of population change, employment, index of economic activities, and waste generation from residential and commercial area should be collected to formulate a good source reduction plan. Communities should be aware about the type and nature of waste material that is disposed off.

Importance of Waste Composition

Realistic goals are set based on the knowledge of composition of the waste stream. The community should be aware of the composition of the waste to get a better idea about the materials to be targeted for the source reduction. Waste audit can be used to determine the composition of a material, product and generating sector. For example, yard waste when burned releases various types of compounds. Grass cycling and back yard composting can be done as a source reduction to minimize the problems associated with yard waste.

Administration and Budget

The scope of Source reduction is very wide when compared to recycling and other disposal options. It involves resource management rather than waste management. It deals with producers, distributers, dealers and consumers. A successful source reduction plan can be drawn based on the decisions about the products and packaging materials used, their composition and the way they are used. Hence source reduction always requires skilled staff with a long term view about the use of materials in the society and changes to be brought for efficient resource utilization with minimum waste generation. Source reduction planning requires an adequate budget like any other conventional waste management option. However, it is not as costly as waste collection and processing. The cost is involved in data collection waste audit, technical assistant, education, equipment and legal issues. The expenditure cannot be incurred in the same budget year. A long term benefit can be attained through source reduction.

Strategies for Source Reduction

Source reduction is accomplished by a number of strategies; however EPA (environmental protection agency) has identified few examples:

- Reducing the amount of packaging material.
- Reducing the amount of product.
- Using packaging material that causes minimum damage/ spoilage to the product.
- Increasing the life span of the product to extend or postpone its disposal through regular maintenance and repairing.
- Reusing existing material, products and packaging material. Example, reconditioning toner cartridges using two sides of paper, reusable bottles etc.
- Modifying or redesigning the product to reduce the quantity and toxicity of the material used.
- Use of alternate materials example, substitute light materials in packing instead of heavier ones.

Few examples of source reduction strategy applied in residential, commercial, institutional, industrial and construction and demolition sources are discussed below.

Residential

According to reports, around 55% to 65% of total municipal solid waste is generated from residential sector. Around 30%of waste generation can be minimized through source reduction practice. The consumer should be aware about the impact they create by generating solid waste and the measures they can undertake to have a positive impact.

- A consumer can purchase products in bulk to minimize the generation of packaging material waste.

- Consumption of locally grown fruits and vegetables will minimize the transportation cost and preservatives used in eatables.
- Purchasing products with high durability and long period of warranty may extend the time of disposal of the product.
- Rechargeable batteries, reusable utensils can save money and waste disposal cost.
- Reusing shopping bags and food containers can also result in source reduction.
- Renting or leasing items can also reduce in waste generation. For example renting tools, lawn movers, leasing computers and many more.
- Using repaired appliances is also an example of green product use.
- Grass cycling returns grass clipping to the lawn and henceforth minimizes the requirement of additional management or disposal.
- Xeriscapping reduces the water usage while landscaping. It also generates less yard waste.

Table: Purchasing strategies for source reduction.

S. No.	Strategy	Examples
1.	Reduce product uses.	Doubled–sided copies. Electronic mails and news.
2.	Rent or lease products or equipment or contract for services.	Rent tools that are need only occasionally. Lease technological products (computer, copiers)
3.	Purchased manufactured, rebuilt, or refurbished products.	Contract services (cleaning, waste removal, printing). Specified rebuilt parts and machinery. Remanufactured laser toner cartridges. Refurbished furniture.
4.	Purchase more durable products.	Long and extended life products (tires and light bulbs. Extended warranties.
5.	Purchase products containing non-hazardous material's.	Green cleaning products. Water based paints and inks. Reduced battery use.
6.	Purchased products that are returnable, reusable, or refillable.	Refillable cups. Reusable utensils and dish wares in cafeterias.
7.	Purchase products in bulk.	Bulk food and concentrates. Supplies in bulk.
8.	Purchase products with less. packaging or reuse packaging.	Purchase concentrate. Avoid purchasing products with secondary packaging.
9.	Share or reuse resources.	Share computers. Reference library.

Large quantities of paper, cardboard, plastics, food waste, electronic waste and other hazardous waste are generated from commercial and institutional sources. Efficient use of paper, onsite food waste composting, use of reusable materials are few of the source reduction initiatives undertak-en in commercial and institutional sources. The source reduction strategies used in commercial and institutional area can be classified into two main categories: changing procurement policies and modifying operations. Changes in procurement policies might reduce the amount of waste generation; encourage manufacturers to develop products and materials that generate less waste; and encourage recycled goods. The policy should be formulated to favour the purchase of more durable, less toxic, reusable, reparable and refillable products. Some options for waste reduction through procurement policy include:

- Suggesting companies to ship goods in reusable containers and take back the packaging material after shipment.
- Purchasing goods with longer warranty period and service contracts.
- Provide preference for using equipment on lease rather than purchasing them.
- Encourage purchasing of items that can reduce use of paper, equipment and computer software that permit double side printing and faxing.
- Providing incentives for reusable refillable and durable equipments.

The measures undertaken for source reduction during operation include:

- Offices with lawn can perform onsite composting of yard waste.
- Educating employees to reduce the use of papers and favour double sided printing and photocopying. Sometimes great reduction in paper usage is achieved by minimizing the number of copies and increasing the intensity of its usage. Alternatives like e- mail, faxing, editing and careful proof reading while printing, necessary printing, and using scrap paper for short memos can also act as strategies for reducing paper waste.
- Using reusable plates and utensils can minimize waste from food service facilities. People can be encouraged by providing discounts for bringing their own plates and utensils. Sometimes they can be charged with additional cost for using disposable items.
- Leasing computers can encourage manufacturers to produce easily upgradable computers. It also increases the responsibilities of the manufacturers to maintain and take back the systems. Sometimes donating computers to schools and non-profit organization can also act as source reduction strategies.
- Reuse and refill packaging containers.
- The packaging materials can be returned back to the company for reuse.
- The product can be redesigned to minimize waste generation.
- Instead of virgin materials other materials collected through material exchange programme can be used.

Table: Consumers strategies for source reduction.

S. No.	Strategy	Examples
1.	Avoid unnecessary packaging.	Chose items with lease or no packaging. Purchase economy size, bulk, or concentrates.

2.	Adopt practices that reduce waste Toxicity.	Use alternative cleaners that do not have hazardous compounds. Use alternative cleaners that do not have hazardous compounds. Choose batteries with reduced mercury. Use digital thermometer instead of Mercury. Use household hazardous waste collection. Use non-toxic inks, dyes and paints.
3.	Conceder reusable products.	Reusable cups, disable and utensils. Cloths napkins and towels instead of Paper. Rechargeable barriers. Refillable detergents.
4.	Maintain and repair durable products.	Choose long lasting and efficient appliances and electronic equipments. Follow proper maintenance scheduled Long lasting tires. Mend and repair clothes, foot ware and bags. Long life florescent light bulbs.
5.	Reuse bags, containers and other items.	Reuse paper, plastics and cloths bags. Reuse scrap paper and envelops. Wash and reuse cans, jugs and containers. Save scrap wood for projects.
6.	Borrow rent or share items.	Rent or borrow parties applies, tools, appliances and electronic equipment, floor and rug cleaners, ladders and many more. Offer before discarding items such as camera and tools. Share newspaper and magazines.
7.	Sell or donate good instead of thronging them out.	Donate clothes, textiles, appliances and furniture to thrift stores and charity. Sell at garbage sales.
8.	Compost yard trimmings and food Scraps.	Backyard composting. Worm composting. Xeriscaping.

Analyzing the waste generated from specific industry may help an individual to redesign the product and manufacturing process to minimize waste generation e.g. Identify that specific process that is involved in waste generation and plan the source reduction strategy accordingly to reduce reuse or avoid waste generating during the process. Most of the time substituting the material for the original one; reducing the weight of the material; marketing a product with minimum or no package; and producing a product that is more durable, reusable and repairable.

Construction and Demolishing

Like municipal solid waste, construction and demolishing waste are increasing in recent times. Source reduction strategies that can be used to minimize construction and demolishing waste include:

- Plan construction in order to meets the requirements.
- Used material with minimum or no toxicity.

- Purchased building materials with minimum packaging.
- Incorporate materials with extended life span.
- Use reusable products.
- Maximize the recovery of material during demolishing e.g. electrical circuits and plumbing material can be recovered and reused or recycled. Doors and windows can be refinished and can be used in new construction.
- Salvaging material can be store in the water house and used in future.

Waste Minimization

Waste minimization involves redesigning products and/or changing societal patterns, concerning consumption and production, of waste generation, to prevent the creation of waste and minimize the toxicity of waste that is produced. The most effective way to reduce waste is by not creating it in the first place. Some examples of waste reduction include using reusable plates and coffee mugs instead of disposable ones, buying durable products that can be repaired rather than replaced, etc.

The Need for Waste Minimization

Waste is not only an environmental problem but also an economic loss as these are resources that are not being used. The amount of waste generated is steadily increasing across the world. This can be attributed to a number of factors like change in consumptive patterns of people, economic growth and globalisation etc. The increasing quantity of waste is in turn putting high demand on land as landfill is one of the commonest techniques for tackling waste. However, globally the available land for this purpose is decreasing due to growing population and demand for land for other purposes. Thus, the shrinking land resource is one of the major drivers for waste minimization.

Secondly, the dumpsites or landfill sites have been identified as a major contributor of greenhouse gases thus confirming their positive role in climate change. This in turn is drawing attention to better waste management practices that emphasize on waste reduction. Thirdly, the cost of waste management is going high. In order to overcome this it becomes essential to target the reduction in waste production, which can be achieved by better production processes. Therefore it can be concluded that the increasing quantity of waste, decreasing availability of landfill site, role of waste in climate change and the high cost of waste management are some of the important drivers for minimizing the quantity of waste.

Benefits of Waste Minimization

While it is obvious that waste minimization supports sound business and economic practices in addition to protecting the environment, other benefits include the following:

- Improved product quality: New technological practices and innovation will not only reduce the generation of waste but also contribute to improved input quality that translates to improved products.

- Economic benefits: Efficiency in product use translates to reduced costs when purchasing materials, thus significantly affecting financial performance.
- The efficiency of production practices: Waste minimization will attain more output of the product for every part of raw material.
- Environmental responsibility: Eliminating or minimizing the generation of waste will make it easy for you to achieve environmental policies, standards and regulations.
- Public image: Embracing waste minimization will boost the reputation of your company, as it is a reflection of proactive movement in the quest to protect the environment.
- Change in RCRA regulatory status: Waste minimization helps a company to change its RCRA regulatory status from a large quantity generator to a small quantity generator or conditionally exempt small quantity generator.
- New customers: People love to buy their goods from companies that are dedicated to improving our environment. Some consumers exclusively seek these organizations out, so you have an opportunity to reach a wider audience. It brings in new customers and generates new money. This is due to your commitment to protecting the environment.
- Fewer Accidents: It reduces the occurrence and likelihood of worker exposure and workplace accidents.
- Protect the environment: Waste minimization increases the efficiency of production. You are able to reduce carbon, air, and water emissions while you work to conserve natural resources. Thus it saves resources and energy, reduces pollution, and diminishes demand for landfill space. All these boost environmental performance.
- Stay ahead of the competition – Reducing waste both in the final product and the manufacturing process, decreases regulatory burdens associated with disposal. More time and money allow you to invest in new ideas and products.

Waste Minimization Techniques

Optimization of Resources

In order to reduce the quantity of waste that is produced by individuals or organizations calls for the optimization of raw materials used in production. For instance, a dressmaker will do well to arrange the pieces of pattern in a certain way along the length of the fabric to use a small portion of the fabric.

Scrap Metal Reuse

Incorporating scraps into the initial stages of manufacturing is a surefire way of ensuring that they do not end up in landfills as waste products. A majority of industries embrace this process, effectively returning rolls that are damaged to the initial production line and in the manufacturing of offcuts, plastic items so that scrap is re-incorporated in the new commodities.

Quality Control Improvement and Process Monitoring

Measures can be put in place to control the number of rejects and ensure it is at a minimum. This may be achieved through increased frequency of inspection as well as increasing the number of inspection points. For instance, the installation of a continuous monitoring device that is automated will help in identifying production problems before they get to an advanced stage.

Exchange of Waste

Here, the waste products from one process are used as raw materials for other processes. Exchange of waste is another means of minimizing waste disposal volumes, especially for waste that may not be eliminated.

Shipping to the Point of use

Here, raw materials as well as other components, are directly delivered at the point of assembly or manufacturing plant ostensibly to minimize handling and use of enclosures and protective wrappings.

Zero Waste

This systems approach is designed to eliminate waste from the source as well as at every point of the supply chain to ensure that no waste is produced. This design philosophy places emphasis on waste prevention and not waste management at the end of the production line.

Reduce the use of Packaging Materials

Start with reducing the use of packaging materials like shrink wrap by redesigning packages to use recyclable or degradable materials. This redesign will not only allow your company to hone in on a newer look and feel but also ensuring your packages do not end up in a landfill. Even if they reach the landfill, you will be happy to know that your material will degrade without causing any harm to the environment.

Reduce Harmful Wastewater

Implement a strategy at your facility to reduce the wastewater and industrial sludge that is a by-product of manufacturing while also reducing the amount of water used during this process since these materials are hard to treat and dispose of. Decreasing the amount of wastewater and sludge will help save money while disposing of the waste properly. Additionally, you will help conserve natural resources and reduce the potential for environmental contamination.

Hold your Employees Accountable

In order to reduce the amount of waste your employees generate, hold your employees accountable to this as well. You can devise a plan that will encourage employees to be environmentally mindful both in and out of the workplace. For example, praise and reward employees that bring their own reusable mugs, cups, plates, and flatware instead of using single-use tableware. In this way, your organization can make the claim that you have made significant steps towards a greener future.

Update your Recycling Program

Another step is to improve your own recycling program. Help employees to get into the habit of properly disposing of materials by creating an internal goal for every month. This goal can be based on the amount of waste your employees or departments recycled, and if they meet the goal, reward them in some way to encourage them to keep up the good work. You can also place specially designed bins in the higher traffic areas and clearly label each one.

Waste Minimization for Households

Households can practice waste minimization by employing various techniques. One of the ways to achieve this is by purchasing adequate sizes and amounts of food. Purchasing large containers of paint when taking small decorating jobs or purchasing large volumes of food than you need will result in wastage. For instance, where cans or packs may be thrown, the remains of the containers should be removed to allow for recycling of the container.

Home composting, thoughtful use of electricity as well as reducing the number of car journeys is also a great way of waste minimization. Generally, buying fewer products or products that last longer, mending worn or broken equipment or clothing, can also minimize household waste. Additionally, households can also minimize wastage of water and cycle or walk to various destinations as opposed to using cars, thereby saving on fuel. Overall, the personal waste reduction will have an effect on the general waste volumes. Consumers may also shun products without eco-labelling.

Waste Minimization in Building Construction

An assessment of streams of waste shows that energy savings may be achieved at minimal cost or no cost within the construction sector. Consequently, the environmental impact of materials may be reduced significantly with reuse.

While at it, it is important to ensure you work with the concerned authorities that include local councils, regional waste authorities, landfill operators or waste recycling contractors. Some of the construction materials that may be recycled include steel, aluminum, gypsum plasterboard, timber, concrete, glass, carpet, plastics as well as bricks and tiles. It is important to put in place waste minimization strategies that have been agreed upon by both parties. A team approach is highly effective in reducing waste.

Assess your Processes

Lastly, always conduct a waste assessment at your facility. This requires examining what waste is generated and how that waste is managed throughout your facility. The technical evaluation also determines whether a proposed Waste Minimization option will work for the specific application. It will help to find new opportunities for waste reduction and cost savings.

Waste Hierarchy

The waste management hierarchy is a concept regarding the waste management and its acts as a base for developing the various waste management strategies. The sole purpose of drafting the hierarchy is to generate the maximum amount of beneficial products from the available wastes.

The most preferred options of the waste hierarchy and will go down towards the least preferred options. The most preferred options are the options having the best effects on the environment, which have been described here in the hierarchical format:

- Avoidance,
- Reduction of Wastes,
- Reuse,
- Recycle,
- Energy Recovery,
- Treatment and Disposal.

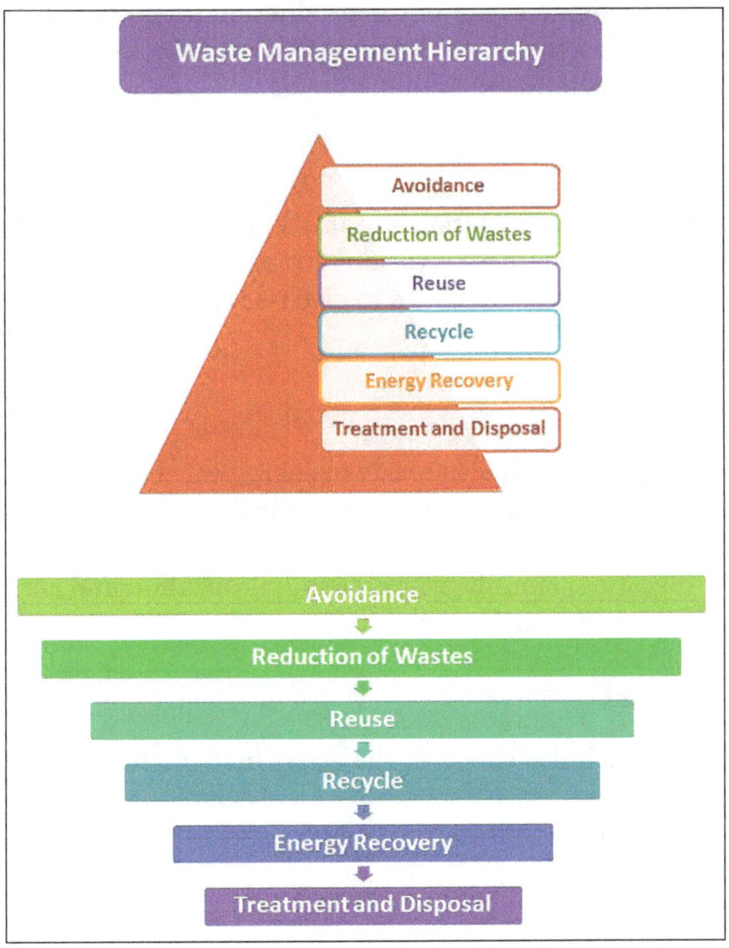

Figure: Waste Management Hierarchy.

Avoidance

This concept focuses on the measures to be taken so as not to create any type of wastes in the first place. This is given the top priority in the waste management program. According to this principle, the manufacturing industries should make use of less hazardous materials in the design and manufacturing of the products. They should develop strategies to have a cleaner and environment friendly production.

Reduction of Wastes

The reduction of wastes or source reduction is the second preferred option of the waste management hierarchy. According to this concept, the companies should take action to make changes in the type of materials that are being used for the production of the specific products, so as to ensure that the by-products are of the least toxicity. This approach helps the industries in various ways. In this way, the industries can be able to effectively save the natural resources.

Reuse

Reuse is another effective Solid waste management strategy, in which the waste is not allowed to enter into the disposal system. The wastes are collected in the middle of the production phase and are again fed along with the source to aid in the production process. This process helps in minimizing the amount of wastes produced as end product, saves the natural resources and reduces the costs associated with the production and manufacturing.

Recycle

In the recycling strategy, the waste materials are implemented in the production of a new product. In this process, the waste materials of various forms are collected and then processed. Post processing, they enter into the production lines to give rise to new products. This process prevents pollution and saves energy.

Energy Recovery

The energy recovery process is also called as waste to energy conversion. In this process; the wastes that cannot be recycled are being converted into useable forms of energy such as heat, light and electricity etc. This helps in the saving of various natural resources. Various processes such as combustion, anaerobic digestion, landfill gas recovery, pyrolization and gasification are being implemented to carry out the conversion process.

Treatment and Disposal

The disposal process holds the last position in the waste management hierarchy. Landfills are the common form of waste disposal.

References

- What-is-the-purpose-of-waste-management: telanganatoday.com, Retrieved 12, July 2020
- General-characteristics-of-waste-management-A-review-272677203: researchgate.net, Retrieved 09, April 2020
- Waste-management-hierarchy: hierarchystructure.com, Retrieved 30, May 2020
- What-is-the-process-of-minimizing-waste: conserve-energy-future.com, Retrieved 10, August 2020
- Source-reduction-strategies: environmentalworks.com, Retrieved 24, May 2020

Chapter 2

Waste Types

There are different types of waste such as hazardous waste, radioactive waste, e-waste, biomedical waste, solid waste and plastic waste. This chapter has been carefully written to provide an easy understanding of these types of wastes.

Hazardous Waste

Wastes are classified as hazardous if they exhibit one or more of ignitability, corrosivity, reactivity, or toxicity. Hazardous wastes are defined as any waste or combination of wastes which pose a substantial present or potential hazard to human health or living organisms because such wastes are non-degradable or persistent in nature or because they can be biologically magnified, or because they can be lethal, or because they may otherwise cause or tend to cause detrimental cumulative effects.

The management of hazardous wastes has become a specialized discipline because of the complex nature of the problem and the solutions available to humanity. The mismanagement examples of hazardous wastes causing disastrous human and environmental consequences are numerous. The management process is based on the definition and classification of the different wastes, and their toxic effects on human and taking in consideration the application of risk management to control human health and environmental impacts of hazardous waste. Hazardous waste management, therefore, deals with minimizing harmful effects on humans and environment by applying special techniques of handling, storage, transportation, treatment, and disposal of hazardous wastes.

Hazardous Characteristics

A useful listing of hazardous characteristics is that provided by the United Nations as part of recommendations relating to the transport of dangerous goods as illustrated in Table.

Table: Hazardous characteristics: Extracted from UN listing.

UN class number	Hazardous characteristic
1	Explosive
3, 4	Flammable
5	Oxidizing
6	Poisonous/infectious
7	Radioactive
8	Corrosive
9	Toxic (delayed or chronic)/ecotoxic

Industrial Wastes

Waste generated from industrial sources can have non-hazardous and hazardous components, with non-hazardous waste usually representing the greater part of the volume. The hazardous component of this waste is relatively small in volume.

This type of waste was identified as hazardous waste when proceeds toxicity test, corrosively test, ignitability test, and some special character test. As a hazardous pollutant, it may impose serious impacts on surrounding environment and such impacts should be quantitatively examined to assess the influence on human health.

Household Waste

Households generate small quantities of hazardous wastes such as oil-based paints, paint thinners, wood preservatives, pesticides, insecticides, household cleaners, used motor oil, antifreeze, and batteries. It has been estimated that household hazardous waste in industrialized countries such as the United States accounts for a total of about 0.5% (by weight) of all waste generated at home, while in most developing countries, the percentage probably is even lower.

Biomedical Waste

There are some of hazardous medical and dental wastes that, when disposed improperly, could cause harm to the environment. It also presents an occupational health hazards to the healthcare personnel who handle these wastes at the point of generation and those involved with their management, that is, segregation, storage, transport, treatment, and disposal.

Healthcare waste that is capable of producing injury or disease including many sorts of hazardous wastes such as:

- Infectious waste: Which contain pathogens namely bacteria, viruses, fungi, or parasites in concentrations sufficient to cause disease in susceptible hosts. Cultures and stocks of infectious agents from laboratory work; tissues and dressing generated from autopsies, surgeries, and treatment of infected patients and animals; materials or equipment in contact with blood and infected body fluids.
- Pathological waste: Including tissue, organs, body parts, human fetuses, and animal carcasses, blood and body fluids.
- Sharps: It comprise syringes, needles, scalpels, saws, infusion sets, knives, blades, broken glass, or other items that can cause cut or puncture wounds.
- Pharmaceutical waste: It covers expired, unused, spilt, and contaminated pharmaceutical products, drugs, vaccines, and sera that are no longer required and need to be disposed of in appropriate manner.
- Genotoxic waste: This type combines cytostatic drugs, vomit, urine, or feces from the patients treated with cytotoxic drugs, chemicals, and radioactive materials. Genotoxic waste has mutagenic, teratogenic, and carcinogenic properties.

- Chemical waste: Discarded solid, liquid, or gaseous chemicals should be considered as hazardous if it is toxic, corrosive, inflammable, or reactive.
- Waste with high content of heavy metals: Mercury (thermometers, blood pressure gauges, amalgam), cadmium (discarded batteries), and lead (reinforced wood panels for radiation proofing in radiology department) generated from hospitals could be represented as a subcategory of hazardous chemical waste.
- Radioactive waste: The use of radioisotopes in vitro analysis of body tissues and fluids, in vivo organ imaging, tumor localization, and treatment and various clinical studies involving certain radionuclides need to be specially managed in a centralized treatment facility for radioactive wastes.

Radioactive Waste

Nuclear applications have been rapidly developed recently, and several nuclear power plants started to work throughout the world. The potential impact of released radioactive contaminants into the environment has received growing attention due to nuclear accidents. Contamination of soil and water by radionuclides due to natural processes, global fallout from nuclear weapon testing, discharges from nuclear installations, disposal of nuclear waste, and occasional nuclear accidents poses serious problems to biological systems. Radioactive waste includes a variety of radionuclides and occurs in a variety of physical and chemical forms. It can be generally classified as low-/intermediate-level radioactive waste and high-level radioactive waste.

Radioactive waste, arising from civilian nuclear activities as well as from weapon activities, poses a potential problem for handling and saving the environment for coming generations. Radioactive waste includes a variety of radionuclides and occurs in a variety of physical and chemical forms. It can be generally classified as low-/intermediate-level radioactive waste and high-level radioactive waste. Nuclear research establishments include, for example, waste containing different organic components, toxic or chemically aggressive constituents, radionuclides with specific properties (high mobility, high chemical activity, volatile elements, etc.), waste difficult for treatment and not appropriate for direct immobilization (e.g., spent organic ion exchange resins and spent liquid scintillation cocktails). For such waste, application of conventional treatment and conditioning options may not be efficient and appropriate in terms of economy, safety, and performance characteristics. In many cases, such wastes are stored awaiting an appropriate treatment and conditioning solution.

The primary sources of radioactive wastes in a country without nuclear fuel cycle activities are nuclear research, production of radioisotopes, application of radioisotopes, and decontamination and decommission of nuclear installations.

Treatment and Disposal

Hazardous Waste Reduction and Minimization

Due to the high potential for public health and environmental damage, some hazardous wastes require special control procedures. Management of these wastes means awareness and control over them from the time of generation through their transportation, temporary storage, treatment and disposal. Effective identification and labeling of wastes by the generators is essential to the effective operation of any manifest-based system. In addition, management of hazardous waste means more than careful disposal. It implies consideration of alternate methods and schemes, both institutional and technical, to hazardous waste reduction. The following measures are commonly adopted as part of Hazardous waste reduction and minimization measures.

- Elimination: The primary priority is to find ways to eliminate uses of hazardous substances altogether. Elimination might be achieved by changing manufacturing processes or by substituting products that can satisfy the same need without creating hazardous wastes (an example would be the substitution of concrete posts in place of, creosote-treated wooden posts).
- Reduction: The next priority is to reduce the amount generated. Again manufacturing processes can be modified to yield substantial waste reduction. For example, changes in the chemical reaction conditions can minimize production of by-product hazardous substances. In some instances, potentially hazardous catalysts, can be replaced by catalysts that are non-hazardous or that can be recycled rather than discarded.
- Recycling: The third strategy is to recycle hazardous substances such as solvents and acids to maximize their use before treatment and disposal becomes necessary. Many industrial processes that use solvents are equipped for solvent recycle for reasons of both economics and pollution control. A number of operations are used in solvent purification. Entrained solids are removed by settling, filtration, or centrifugation. Drying agents may be used to remove water. Fractional distillation is used to separate solvent from impurities, water and other solvents.
- Finally, hazardous substances can be treated to reduce their volume and toxicity.

Eliminating, reducing, and recycling hazardous substances reduces treatment and disposal costs and often the generators can find it in their own economic best interest to actively seek out these source reduction opportunities as a way to reduce the overall cost of doing business.

Hazardous Waste Treatment Technologies

The selection of a treatment process for a waste stream depends on among other factors the nature of the waste, desired characteristics of the output stream. Even with after vigorous hazardous waste reduction program, there will still be large quantities of hazardous wastes that will require treatment and disposal. The treatment technologies have been categorized as physical, chemical, biological, thermal, or stabilization/fixation.

- Physical treatment processes include gravity separation, phase change systems, such as air and steam stripping of volatiles from liquid wastes, and various filtering operations, including carbon adsorption.
- Chemical treatment transforms waste into less hazardous substances using such techniques as pH neutralization, oxidation or reduction, and precipitation. Biological treatment uses microorganisms to degrade organic compounds in the waste stream.

- Thermal destruction processes include incineration, which is increasingly becoming a preferred option for the treatment of hazardous wastes, and pyrolysis, which is the chemical decomposition of waste is brought about by heating the material in the absence of oxygen.
- Stabilization techniques involve removal of excess of water from a waste and solidifying the remainder either by mixing it with a stabilizing agent such as Portland cement, or vitrifying it to a glassy substance.

Most treatment measures have both physical and chemical aspects. The appropriate treatment technology for the hazardous wastes depends on the nature of the wastes. The type of physical treatment to be applied to wastes depends strongly upon the physical properties of the material treated, including the state of matter, solubility in water and organic solvents, density, volatility, boiling point and melting point.

Physical Treatment Methods

Adsorption

Adsorption on activated carbon occurs when a molecule is brought up to its surface and held there by physical and /or chemical forces. This process is reversible, thus allowing activated carbon to be regenerated and reused by proper application of heat and steam, or solvent. The factors that relate to adsorption capacity are:

- Greater surface area produces greater adsorption capacity (Eg: Activated carbon has large surface area (500-1500 m^2/g)].
- Adsorptivity increases as the solubility of the solute (in solvent) decreases. Thus, for hydrocarbons, adsorption increases with molecular weight.
- For solutes with ionizable groups, maximum adsorption will be achieved at a pH corresponding to minimum ionization.
- Adsorption capacity decreases with increasing temperature.

One additional point to be noted is that biological activity usually takes place in a carbon bed. If the concentration of the adsorbed species is high enough and the material is biodegradable and nontoxic to the bacteria, then biological activity may significantly increase the effective removal capacity. Removal through adsorption by activated carbon has been applied to non-aqueous waste stream such as petroleum fraction, syrups, vegetable oils, and pharmaceutical preparations. Color removal is the most common objective in such cases. Current waste treatment applications are limited to aqueous solutions.

Resin Adsorption

Waste treatment by resin involves two basic steps: (1) contacting the liquid waste stream with resin and allowing the resin to adsorb the solutes from the solution; and (2) subsequently regenerating the resins by removing the adsorbed chemicals, by simply washing with proper solvent.

The adsorption of a nonpolar molecule on to a hydrophobic resin (e.g., styrene divinyl-benzene-based resin) results primarily from the effect of Vander Waal's forces. In other cases, other type of interactions such as dipole-dipole interaction and hydrogen bonding are important. In a few cases ion exchange mechanism may be involved. For the removal of organic dye wastes from water, two different resins were employed: In this case the waste stream is first contacted with a normal polymeric adsorbent and then with an ion exchange resin.

Sedimentation

Sedimentation is a physical process whereby particles suspended in a liquid settle by means of gravity. The fundamental elements of most sedimentation processes are:

- A basin or container of sufficient size to maintain the liquid to be treated in a relatively quiescent state for a specified period of time.
- A means of directing the liquid to be treated into the above basin in a manner conducive to settling.
- A means of physically removing the settled particles from the liquid.

Sedimentation can be either a batch or a continuous process. Continuous processes are by far the most common, particularly when large volumes of liquid are to be treated. This technique has been widely used in the removal of heavy metals from iron and steel industry wastewater; removal of fluoride from aluminium production wastewater; and removal of heavy metals from wastewater from copper smelting and from metal finishing industry and wastewater stream from organic chemicals.

Electrodialysis

The electrodialysis involves the separation of an aqueous stream (more concentrated in electrolyte than the original) and a depleted stream. Success of the process depends on special synthetic membranes, usually based on ion exchange resins, which are permeable only to a single type of ion. Cation exchange membranes permit passage only of positive ions, under the influence of electric field, while anion exchange membranes permit passage only of negatively charged ions.

The feed water is passed through compartments formed by the spaces between alternating cationpermeable and anion permeable membranes held in a stack. At each end of the stack is an electrode having the same area as the membranes. A dc potential applied across the stack causes the positive and negative ions to migrate in opposite directions. This technique has been used for desalination to produce potable water from brackish well water. In food industry electrodialysis is used for desalting whey. The chemical industry uses this technique for enriching or depleting solutions, and for removing mineral constituents from product streams.

Reverse Osmosis

This technique which is most widely used consists of a membrane permeable to solvent but impermeable to most dissolved species, both organic and inorganic. These devices use pressure to force the contaminated water against the semipermeable membrane. The membrane acts as a filter, allowing the water to be pushed through the pores, but restricting the passage of larger molecules that are to be removed.

Cellulose acetate membranes were used in the past, but nowadays polysulphones and polyamides are increasingly popular for use at high pH values. Because of the susceptibility of the membranes to chemical attack and fouling, and the susceptibility of the flow system to plugging and erosion, it is common to pre-process the feed water to remove oxidising materials. The reverse osmosis technique has been widely used for desalination of sea or brackish water. It has also been successfully used in the treatment of electroplating rinse waters, not only to meet effluent discharge standards, but also to recover concentrated metal salt solutions for reuse. It has also been used for treatment of waste stream from paper and food processing industries.

Solvent Extraction

Solvent extraction is the separation of the constituents of a liquid solution by contact with another immiscible liquid. If the substances comprising the original solution distribute themselves differently between the two liquid phases, a certain degree of separation will result and this may be enhanced by the use of multiple contacts.

The major application of solvent extraction to waste treatment has been in the removal of phenol from by-product water produced in coal coking, petroleum refining, and chemical synthesis that involve phenol.

The use of supercritical fluids (SCFs) most commonly CO_2 as extraction solvent, has been one of the more promising approaches to solvent extraction. SCFs are fluids existing at or above the lowest temperature at which condensation may occur. Above the critical temperature certain fluids exhibit characteristics that enhance their solvent properties. Organic materials, which are only slightly soluble in particular solvents at room temperature, become completely miscible with the solvent when under supercritical conditions. The excellent solvent properties result from the rapid mass transfer ability and the very low density that characterizes an SCF. Major advantages of SCFs are short residence times with no char formation. Some of the important application of these SCFs, has been in the extraction of organohalide pesticide from soil, extraction of oil from emulsions used in aluminium and steel processing, and regeneration of spent activated carbon. Supercritical ethane has been used to purify waste oils contaminated with PCBs, metals and water.

Distillation

Distillation is expensive and energy intensive and can probably be justified only in cases where valuable product recovery is feasible (e.g., solvent recovery). This technique has only limited application in the treatment of dilute aqueous hazardous wastes.

Evaporation

Evaporation process is used for the treatment of hazardous waste such as radioactive liquids and sludges and concentrating of plating and paint solvent waste among many other applications. It is capable of handling liquids, slurries and sometimes sludges, both organic and inorganic, containing suspended or dissolved solids or dissolved liquids, where one of the components is essentially non-volatile. It can be used to reduce waste volume prior to land fill disposal or incineration.

The major disadvantages of evaporation are high capital and operating costs and high energy requirements. This process is more adaptable to waste waters with high concentrations of pollutants.

Filtration

Filtration is well-developed economical process used in the full scale treatment of many industrial waste waters and waste sludges. Energy requirements are relatively low, and operational parameters are well defined. However, it is not a primary treatment process and is often used in conjunction with precipitation, flocculation, and sedimentation to remove these solids.

Flocculation

The various phenomena that occur during flocculation can be grouped in to two sequential mechanisms.

- Chemically induced destabilization of repulsive surface related forces, thus allowing particles to stick together when they touch.
- Chemical bridging and physical enmeshment between the non-repelling particles, allowing for the formation of large particles.

Chemicals used for flocculation include alum, lime, ferric chloride, ferrous sulphate and poly electrolytes. Poly electrolytes consist of long chain, water soluble polymers such as polyacrylamides. They are used either in conjunction with inorganic flocculants, or as primary flocculating agent. The inorganic flocculants such as alum, upon mixing with water, the slightly higher pH of water causes them to hydrolyze to form gelatinous precipitates of aluminium hydroxide. It is partially because of their large surface area, they are able to enmesh small particles, and thereby create larger particles.

Once suspended particles have been flocculated into larger particles, they usually can be removed from the liquid by sedimentation, provided that a sufficient density difference exists between the suspended matter and the liquid.

Chemical Treatment Methods

Chemical Oxidation and Reduction

Oxidation reduction is one important route used for the treatment of hazardous wastes. It is a process by which the hazardous wastes are converted into a form that shows reduced or no toxicity, reduced or no mobility and remains inert. The process can be short or medium term technology and uses oxidizing agents for the treatment of hazardous waste. Few examples of oxidizing agents include ozone, hydrogen peroxides, chlorine, chlorine dioxide and others. Some examples of oxidation reduction treatment of hazardous wastes are given below. Chlorine in alkali solution is used, which converts the CN^- to cyanate (less toxic form). More addition of chlorine oxidizes the cyanate to CO_2 and N_2.

$$NaCN + Cl_2 + 2NaOH \rightarrow NaCNO + 2NaCl + H_2O$$

$$2NaCNO + 3Cl_2 + 4NaOH \rightarrow 2CO_2 + N_2 + 6NaCl + 2H_2O$$

Hexavalent chromium Cr(VI) is reduced to trivalent chromium Cr(III) by using Sulphur dioxide as reducing agent:

$$3SO_2 + 3H_2O \rightarrow 3H_2SO_3$$

$$2CrO_3 + 3H_2SO_3 \rightarrow Cr_2(SO_4)_3 + 3H_2O$$

Iron (II) is oxidize and precipitated as ferric hydroxide:

$$4Fe^{2+} + O_2 + 10H_2O \rightarrow 4Fe(OH)_3 + 8H+$$

Sulphur dioxide is oxidized to sulphuric acid:

$$2SO_2 + O_2 + 2H_2O \rightarrow 2H_2SO_4$$

Ozonolysis

Ozone is used to oxidize a wide variety of contaminants present in waste water and sludge. Ozone is generated by an electrical discharge through dry air or oxygen.

Ozone is also used to destruct TCDD and PCBs. TCDD demonstrated that if the dioxins were suspended as an aerosol in combination with CCl_4, degradation of 97% TCDD would be possible. Destruction of polychlorinated phenols and pesticides by ozone can be done in conjunction with UV radiation. In both the cases the key requirements were to concentrate the TCDD in a medium where they were susceptible to attack and provide a free radical for reaction with dioxin molecule.

$$(CH_2O) + 2[O] \rightarrow CO_2 + H_2O$$

$$CH_3CHO + [O] \rightarrow CH_3\ COOH$$

Acid-Base Neutralization

A waste showing a pH less than 2 or more than 12.5 are categorized as corrosive hazardous wastes. They can be neutralized by acid base reactions. Lime is used to neutralize acid wastes. Addition of slaked lime [$Ca(OH)_2$] to acidic wastes in a continuously stirred chemical reactor results in neutralization of hazardous waste. Lime is commonly preferred due to its economic nature and optimum solubility. Due to limited solubility, excess lime does not increase the pH values. Similar to lime, sulphuric acid and acetic acid is used for neutralizing alkaline wastes. Though sulphuric acid is inexpensive, at certain times acetic acid is preferred as it is nontoxic and biodegradable. Certain alkaline wastes are also neutralized by bubbling gaseous CO_2 forming carbonic acid. CO_2 is easily available and can be collected from the exhaust gas of any combustion process. In addition to this, may other waste treatment processes require prior pH adjustment. The process includes oxidation/reduction, adsorption, wet air oxidation, ion-exchange, stripping and biochemical treatment.

Chemical Precipitation

Precipitation is applicable for mixed wastes that contain both organic and inorganic species. By this method the toxic substances or metals are precipitated using chemicals and removed from the solution by sedimentation or settling or filtration. Alteration of solution pH will minimize the solubilization of heavy metals which can further be precipitated as precipitable hazardous constituent. Slaked lime [$Ca(OH)_2$] or caustic soda is used for precipitation of the metal ions as metal hydroxides. For example, the following reaction suggests the use of lime to precipitate the metal as hydroxide:

$$M_2 + Ca(OH)_2 \leftrightarrow M(OH)_2 + Ca^{2++}$$

Chromium is precipitated as hydroxide:

$$Cr^{3+} + 3(OH^-) \rightarrow Cr(OH)_3$$

Sodium carbonate is also used to precipitate metals as hydroxides. ($Fe(OH)_3$. XH_2O), carbonates ($CdCO_3$), basic carbonate salts ($2PbCO_3.Pb(OH)_2$).

Carbonate ion hydrolyses in water to give hydroxide ion:

$$CO3_2^- + H_2O \rightarrow HCO_3^- + OH^-$$

Metals can be precipitated as sulphides from solution. Precipitating the metals as sulphides will facilitate removal of low concentration of metals. Ferrous sulphide is popularly used. Sodium borohydride is another reducing agent that precipitates metal in elemental form:

$$4Cu^{2+} + NaBH_4 + 2H_2O \rightarrow 4Cu + NaBO_2 + 8H^+$$

Hydrolysis

Hydrolysis treatment can be given to those hazardous waste constituents which shows high reactivity with water. Examples of those substances are halides, carbide, hydride, alkoxide, and active metal:

$$SiCl_4 + 2H_2O \rightarrow SiO_2 + 4HCl$$

$$CaC_2 + 2H_2O \rightarrow Ca(OH)_2 + C_2H_2$$

$$NaAlH_4 + 4H_2O \rightarrow 4H_2 + NaOH + Al(OH)_3$$

$$NaOC_2H_5 + H_2O \rightarrow NaOH + C_2H_5OH$$

$$Ca + 2H_2O \rightarrow Ca(OH)_2 + H_2$$

Ion Exchange

Ion exchange is best suited for wastes that have undergone pre-treatment or minimum treatment with other chemical and physical methods. It is suitable for removing inorganic chemical species that are in the dissolved form unlike precipitation, flocculation or sedimentation. Basically it is used as a final polishing step to reduce an inorganic species that could not be reduced to satisfactory levels by preceding treatment processes. Nickel cyanide complex and chromate ions from hazardous waste solutions are removed by anion ion exchangers.

$$2Res\ OH^- + [Ni(CN)_4]^{2-} \rightarrow (Res^+)_2\ [Ni(CN)_4]^{2-} + 2OH^-$$

$$2Res\ OH^- + CrO_4^{2-} \rightarrow (Res^+)_2\ (CrO_4^{2-}) + 2OH^-$$

Ion exchange resins have also been used in the removal of radionuclides from radioactive wastes.

Thermal Treatment Methods

Thermal destruction of hazardous waste is done by incineration and pyrolysis. Incineration is a process that combusts the wastes at high-temperature (900-1100 °C) in the presence of oxygen. This process converts waste to an inert, less toxic and less bulky material namely, the ash. Ash is an non-combustible inorganic residue. Incineration reduces the volume of waste by 90%. The organic elements such as carbon, hydrogen and oxygen are converted wholly or partially to gaseous form. The principal products of incineration are CO_2, water vapor and ash.

$$C(organic) + O_2 \rightarrow CO_2 + heat$$

$$4H(organic) + O_2 \rightarrow 2H_2O + heat$$

The heat generated during the incineration process is used for destruction of organochlorine hazardous wastes. The heat cleaves the C-Cl bonds of organochlorine compounds. If the organochlorines are inflammable, then supplementary fuel such as methane or petroleum-liquid is used. Though incineration process comes with many advantages, still its application is limited due to the release of by-products of hazardous waste incineration such as sulphur, nitrogen, halogen and heavy metals (mercury, arsenic, selenium, lead and cadmium). This results in problems of air pollution. To overcome the air pollution problem, the incinerators need to be fitted with air pollution control equipment. Temperature, Availability of sufficient oxygen and ample residence time are the critical factors governing incineration process. Destruction removal efficiency (DRE) is used to assess the efficiency of hazardous-waste incinerators. DRE can be calculated as the percentage of mass difference of input(feed) and output (stack emission) waste constituents through the incinerator. DRE has to be calculated separately for different compound/ constituent. According to RCRA (Resource conservation and Recovery Act) requirement, a minimum DRE of 99.99 % for most of organic compounds and a DRE of 99.9999 % for dioxins and dibenzo furans should be achieved for efficient combustion.

Advantages of Incineration

- Reduces 90% volume of waste.
- Can be scaled to handle large volumes of liquid waste.
- It is the best method for "mixed wastes" and biologically hazardous wastes.
- Does not require large area of land.
- Compact units.

Disadvantages of Incineration

- Capital and operation cost is very high.
- The ash must be disposed of properly.
- The gaseous and particulate products of combustion should be controlled by air pollution control devices.

Wet Air Oxidation

Wet air oxidation is the process by which the dissolved or suspended organic substances are oxidized in aqueous phase at high temperature and pressure of 150-325 °C and 2000 kPa to 20,000 kPa, respectively. Water acts as a catalyzing agent and moderator by removing excess heat in the reaction mixture. Water also acts as an excellent heat transfer medium so that the wet air oxidation process becomes thermally self-sustaining at low organic feed conditions. In wet air oxidation, the waste is pumped into the system through a heat exchanger into the reactor with high-pressure pump and mixed with air from an air compressor. High pressure facilitates dissolving of oxygen in water. Within the reactor, the organic waste is reacted with atmospheric oxygen. Sometimes the oxidation is carried out in the presence of catalysts. Cyanide from electroplating waste solutions are removed by wet air oxidation process. The process can be used for the removal of:

$$2Na^+ + 2CN^- + O_2 + 4H_2O \rightarrow 2Na^+ + 2HCO_3^- + 2NH_3$$

Photolysis

In photolysis, light is used as an agent to cleave the chemical bonds of hazardous waste materials. Initially, the chemical species is excited by light. It is usually a diradical state because diradical possess sufficient energy to undergo chemical reaction. The success or effectiveness of the process depends on the chemical structure of the compound, efficiency of radiation source and medium in which the reaction is carried out. The end product also determines the efficiency of the process. Polychlorinated biphenyls (PCBs) and chlorinated dibenzo-p-dioxins(CDDs) are destroyed by photolysis. The three requirements of photolysis of tetrachloro dibenzo-p-dioxin (TCDD) are 1) Dissolution in a light transmitting film 2) presence of organic hydrogen odor and 3) ultraviolet light. Generally, during photolysis reactions, HO• is formed which plays an important role in chain reactions resulting in destruction of the compound.

Biological Treatment of Hazardous Wastes

Biological processes are a simple, cost effective and eco-friendly process. However its application to hazardous waste is slightly limited due to toxic nature of the waste which is harmful to the biological agents. The biodegradability of the waste depends on the physical and chemical properties of the compound. Selection of appropriate organism and optimum conditions can favor degradation of chemical compounds that are considered harmful or biocidal. Eg. Phenol and phenolic compounds can be treated by proper acclimatization of microorganisms to the hazardous waste. Likewise, biorefractory compounds eg. DDT and PCBs are also degraded by bacteria. DDT is degraded by Pseudomonas and PCB by anaerobic bacteria. According to studies, optimum pH and oxygen levels can result in better degradation of waste. Both aerobic and anaerobic bacteria are used for hazardous waste biodegradation. The presence of inorganic substances limit the biodegradation of waste as they inhibit enzymatic activity and further get adsorbed on to the microbial cell coating. Landfill leachate is treated with microorganisms. During the process H_2S is generated which precipitates toxic heavy metal ions as their respective sulphides. The overall degradation of the hypothetical organic compound (CH_2O) can be written as follows:

$$2(CH_2O) \rightarrow CO_2 + H_2O$$

Land Treatment

The application of waste to soil surface in a controlled manner is termed as land treatment of hazardous wastes. This process facilitates containment of waste followed by biological and chemical degradation of organic waste. This type of treatment is different from landfilling. Here, the soil is used to detoxify, immobilize, and degrade all or a portion of the applied waste. Landfills contain the waste and control its migration from the site. Hazardous wastes should not be placed in a land treatment site unless the waste can be made less hazardous or non-hazardous by the reactions occurring in the soil. The waste is added or mixed to the aerobic soil horizon in a controlled manner at appropriate conditions. By this process, the physical, chemical and biological characteristics of the waste are altered. Bacteria, fungi and actinomycetes plays a major role in land treatment. Genus Agrobacterium, Anthrobacteri, Bacillus, Flavobacterium, Pseudomonas are some of the bacteria involved in degradation. Wastes from petroleum refining, biodegradable organic chemical wastes including organochlorine compounds are suitable for land treatment. Land treatment should not be applied to waste containing acids, bases, toxic inorganic compounds, salts, heavy metals and excessively soluble volatile and flammable organic compounds.

Land Disposal

After treatment, the hazardous wastes are converted into suitable form for disposal. Immobilization, stabilization, fixation, and solidification are some of the popularly used techniques for preparation of waste for disposal. Land disposal techniques include Landfills, surface impoundments and underground injection wells.

Landfills

In order to protect public health and environment the design of hazardous waste landfills should be adequate. Figure shows the typical cross section of a hazardous waste landfill.

The liner system, waste emplacement and cover system are the three components of landfills. Liners should be impermeable and avoid the contact between waste and the surrounding environment. Leachate collection system, drains and leak detectors too form an integral part of the landfill system. They ensure that leachate does not leave the disposal site. Hazardous wastes are contained in secured landfills provided with double liner system. The liner system consists of natural (clay), synthetic material (polymers), and a leachate collection. The primary barrier is a synthetic polymeric flexible membrane with a separation layer of thin geotextile or polymeric material (HDPE, LDPE, PVC etc). Secondary barrier is composed of clay, bentonite enhanced soil or geosynthetic clay. A leachate collection tubes are layered between the waste and the protection layer. It channelizes the leachate to a pumping station where they are removed and taken up for further treatment.

An intermediate high permeability drainage layer is incorporated between primary and secondary barrier. This layer is for the collection of leachate and gas. Every barrier is separated by geotextile fabric. The landfill is sloped to avoid infiltration of water and run off associated with it. The leachate that is collected is treated by (1) passing it through a column containing sorbent material such as carbon or flyash. (2) subjected to suitable physical-chemical units such as chemical addition, flocculation, sedimentation, pressure filtration, pH adjustment, and reverse osmosis to remove the dissolved waste. The secondary leachate collection system acts as back-up system which will be useful during failure of primary leachate system. Continuous monitoring of ground water from the wells will help us to determine the failure of primary and secondary liner system.

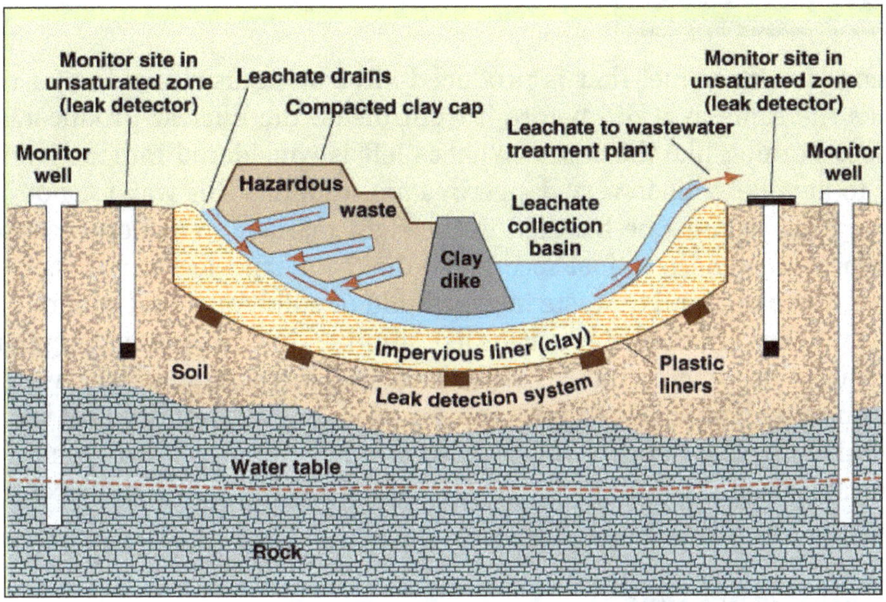

Figure: Schematic of hazardous waste landfill.

Waste is loaded inside the landfill after preliminary treatment (i.e) solidification, stabilization etc. Th cover or capping is also essential component of the waste landfill. The purpose of the landfill cover is to protect the waste, prevent the entry of rain and surface water into the waste, control the release of landfill gas, to prevent ingress of air and enable the growth of short rooted plant when covered with top soil. Landfills are allowed with sufficient vent points so that if methane is generated, it may be burned off continuously.

Surface Impoundments

A surface impoundment is a man-made excavation, diked area, or natural topographic depression designed to hold an accumulation of liquid wastes. The construction is similar to landfill where the bottom walls function like a liner which is impermeable to liquids. The surface impoundments are also provided with provision for leachate collection. Geological siting and selection of low permeable material like soil and clay are very important as they will arrest or minimize the chances of pollution.

Underground Injection

Underground injection or deep-well disposal is a method by which the hazardous waste materials are injected into underground strata under pressure. The underground strata must be impermeable and separated from the aquifers. During this process, the chance of chemical reactions with waste constituent and mineral is possible. This will result in corrosion, clogging problems. The main concern with underground injection is the potential for contaminating underground drinking water supplies, if the disposal well is not properly cased or if it is damaged.

Radioactive Waste

Radioactive waste is nuclear fuel that is produced after being used inside of a nuclear reactor. Although it looks the same as it did before it went inside the nuclear producer, it has changed compounds and is nothing like the same. What is left is considered radioactive material and is very dangerous to anyone. This is very dangerous and remains this way for not just a few years but thousands of years. It must be handled in the right manner so as not to cause a ton of devastation in the world. It could take just seconds to die from exposure to radioactive materials. In short, radioactive waste is a kind of waste in gas, liquid or solid form that contains the radioactive nuclear substance. There are many industries like mining, defense, medicine, scientific research, nuclear power generation which produce by-products that include radioactive waste. The radioactive waste can remain radioactive for a few months, years or even hundreds of years and the level of radioactivity can vary. The radioactive waste is extremely toxic as it can remain radioactive for so long and can cause acute radiation sickness when it first comes out of the reactor if you stood within a few meters of it while it was unshielded.

Sources of Radioactive Waste

Radioactive waste is generated from a variety of industries. Any waste that contains or contaminated with radioactive materials are called as radioactive waste. They can be generated from operation and decommissioning of nuclear facilities, radionuclides using industry, radionuclides containg raw material, medicine and research. Radioactive waste is also generated during the cleanup of sites affected by radioactive residues from various operations or accidents. At certain times, the materials generated as waste may be reused or recycled.

Radioactive Waste Arises from Many Different Activities

- Operation and decommissioning of nuclear facilities (e.g. nuclear power plants).
- Application of radionuclides in industry, medicine, and research.
- Cleanup of contaminated sites.
- Processing of raw materials containing naturally occurring radionuclides.

Radioactive waste from the nuclear fuel cycle consists of materials with varied range of radioactivity levels. The spent nuclear fuel contains radionuclides like uranium, plutonium, their decay products and many fission products. In many countries, the spent fuel is reprocessed to separate U and

Pu. Handling of spent fuel should be done with great care through sophisticated remote handling techniques. Proper shielding should be provided while handling these wastes.

Operational waste includes a variety of lower-level wastes, either in the form of filtering residues (e.g. ion exchange resins) or as contaminated supplies (e.g. clothing) containing fission and activation products.

Decommissioning of nuclear power stations gives rise to large volumes of contaminated material, including some highly-activated reactor components as well as structural materials that may be slightly contaminated by activation products and/or fission products.

Wastes generated from other units of the nuclear fuel cycle include milling wastes containing decay products of natural uranium, such as radium and radon and radon daughters etc. Long lifetimes of these materials pose a great problem. Sometimes, the radioactivity is increased when few daughter products are combined. Facilities where the ore is refined, enriched and processed into fuel also result in radioactive waste. In the case of depleted uranium tails, in addition to direct toxicity of the uranium, daughter products (radium and radon) buildup on long time scales also increase the toxicity. e. Radionuclides are used extensively in industrial research, agricultural research, geological exploration, constructions for measuring thickness, density or volume of materials. They are also used in smoke detectors; tracers; sealed sources for irradiation, heat/ power source.

Classification of Radioactive Waste

The classification scheme of radioactive waste has been modified with the intention to address the shortcomings identified with the previous scheme and to reflect experience gained in developing, operating and assessing the safety of disposal facilities. The classification scheme developed previously is not completely comprehensive in that it does not cover all types of radioactive waste, nor does it provide a direct linkage with disposal options for all types of radioactive waste. These aspects of the former classification scheme have been deemed limitations on its use and application.

A comprehensive range of waste classes has been defined and general boundary conditions between the classes are provided. More detailed quantitative boundaries that take into account a broader range of parameters may be developed in accordance with national programs and requirements. In cases when there is more than one disposal facility in a State, the quantitative boundaries between the classes for different disposal facilities may differ in accordance with scenarios, geological and technical parameters and other parameters that are relevant to the site specific safety assessment.

A number of elements of the previously set out classification scheme have been retained. A complete range of waste classes has been defined and general boundary conditions between the classes are provided to reflect the experience gained in developing, operating and assessing the safety of disposal facilities. More detailed quantitative boundaries will be prepared according to the national programmes and requirements. In cases when there is more than one disposal facility in a State, the quantitative boundaries between the classes for different disposal facilities may differ in accordance with scenarios, geological and technical parameters and other parameters that are relevant to the site specific safety assessment. Six classes of waste have been derived and used as the basis for the classification scheme:

- Exempt waste (EW): Waste that meets the criteria for clearance, exemption or exclusion from regulatory control for radiation protection purposes.
- Very short lived waste (VSLW): Waste that can be stored for decay over a limited period of up to a few years and subsequently cleared from regulatory control according to arrangements approved by the regulatory body, for uncontrolled disposal, use or discharge. Eg. Waste containing primarily radionuclides with very short half-lives often used for research and medical purposes.
- Very low level waste (VLLW): Waste that does not necessarily meet the criteria of EW, but that does not need a high level of containment and isolation and, therefore, is suitable for disposal in near surface landfill type facilities with limited regulatory control. Such landfill type facilities may also contain other hazardous waste. Typical waste in this class includes soil and rubble with low levels of activity concentration. Concentrations of longer lived radionuclides in VLLW are generally very limited.
- Low level waste (LLW): Waste that is above clearance levels, but with limited amounts of long lived radionuclides are categorized as LLW. Such waste requires robust isolation and containment for periods of up to a few hundred years and is suitable for disposal in engineered near surface facilities.

This class covers a very broad range of waste. LLW may include short lived radionuclides at higher levels of activity concentration, and also long lived radionuclides with relatively low levels of activity concentration. Low level Waste (LLW) is generated from hospitals and industry, as well as the nuclear fuel cycle. It comprises paper, rags, tools, clothing, and filters etc. which contain small amounts of mostly short-lived radioactivity. It does not require shielding during handling and transport and is suitable for shallow land burial. To reduce its volume, it is often compacted or incinerated before disposal.

LLW includes everything not included in one of the other categories. LLW is not necessarily less radioactive than HLW and may have higher specific activity in becquerel per gram. The distinguishing feature of LLW is that it contains virtually no alpha emitters. To ensure appropriate disposal, the NRC has designated several classes of LLW:

- Class A contains only short-lived radionuclides or extremely low concentrations of longer-lived radionuclides, and must be chemically stable. It is disposed of in designated LLW landfills as long as it is not mixed with hazardous or flammable waste.
- Class B contains higher levels of radioactivity and must be physically stabilized before transportation or disposal. It must not contain free liquid.
- Class C is waste that will not decay to acceptable levels in 100 years and must be isolated for 300 years or more. Power plant LLW is in this category.
- Greater than Class C (GTCC) will not decay to acceptable levels in 300 years. A small fraction of power plant Class C waste is in this category. Some nations like Sweden treat GTCC waste like HLW, and the United States may do the same.

Mixed Low-Level Waste (MLLW) is LLW that contains hazardous waste as defined under the Resource Conservation and Recovery Act (RCRA).

- Intermediate level waste (ILW): Waste that, because of its content, particularly of long lived radionuclides, requires a greater degree of containment and isolation than that provided by near surface disposal. However, ILW needs no provision, or only limited provision, for heat dissipation during its storage and disposal. Intermediate level Waste (ILW) contains higher amounts of radioactivity and some requires shielding. It typically comprises resins, chemical sludges and metal fuel cladding, as well as contaminated materials from reactor decommissioning. It may be solidified in concrete or bitumen for disposal. ILW may contain long lived radionuclides, in particular, alpha emitting radionuclides that will not decay to a level of activity concentration acceptable for near surface disposal during the time for which institutional controls can be relied upon. Therefore, waste in this class requires disposal at greater depths, of the order of tens of meters to a few hundred meters.
- High level waste (HLW): Waste with levels of activity high enough to generate significant quantities of heat by the radioactive decay process or waste with large amounts of long lived radionuclides that need to be considered in the design of a disposal facility for such waste. A limit of 400 Bq/g on average (and up to 4000 Bq/g for individual packages) for long lived alpha emitting radionuclides has been adopted in some States. For long lived beta and/or gamma emitting radionuclides, such as 14C, 36Cl, 63Ni, 93Zr, 94Nb, 99Tc and 129I, the allowable average activity concentrations may be considerably higher (up to tens of kilo becquerel per gram) and may be specific to the site and disposal facility. Disposal in deep, stable geological formations usually several hundred meters or more below the surface is the generally recognized option for disposal of HLW. High level Waste (HLW) arises from the use of uranium fuel in a nuclear reactor and nuclear weapons processing. It contains the fission products and transuranic elements generated in the reactor core. It is highly radioactive and hot. HLW accounts for over 95% of the total radioactivity produced in the process of nuclear electricity generation.

Table: Common Categories of Radioactive Waste.

Waste Category	Description of Waste Category	Common Sources of Waste	Common Radionuclides in Waste and their Half-Life (y=years)
High-level Radioactive Waste	Highly radioactive material that is deemed a waste that requires special precautions by humans, including remote handling and use of shielding; also includes spent fuel and waste resulting from the reprocessing of used fuel.	Partially used fuel from nuclear power reactors; liquid waste from the reprocessing of spent fuel taking place outside the United States.	strontium-90 half-life: 29.78 y cesium-137 half-life: 30.07 y
Transuranic Waste	Material that is deemed a waste that contains radionuclides with an atomic number greater than that of uranium.	Weapons-production waste included mixed transuranic waste.	plutonium-238 halflife: 87.7 y americium-241 halflife: 432.7 y

Mixed Waste	Material that is deemed a waste that contains both radionuclides and a characteristic or listed hazardous waste.	Weapons-production waste and some research wastes.	plutonium-239 halflife: 24,100 y plutonium-241 halflife: 14.4 y
Naturally Occurring Radioactive Material (NORM) Waste	Material that is deemed a waste that contains radionuclides that are present on Earth without any human interaction.	Scale buildup on pipe walls that carry petroleum products.	radium-226 half-life: 1,599 y radium-228 half-life: 5.76 y
Uranium or Thorium Mill Tailings Waste	The tailings material created as a byproduct by the extraction of uranium or thorium from natural ore formations.	Production exclusively at the site of milling for rare earth extraction.	radium-226 half-life: 1,599 y thorium-230 half-life: 75,400 y
Low-Level Radioactive Waste (LLRW) (and Intermediate Waste outside U.S.)	Material that is deemed a waste that generally has been contaminated by or contains short-lived radionuclides or longer-lived radionuclides in relatively low concentrations. Lowlevel radioactive waste is further segregated into classes.	Industrial trash from nuclear power plants; medical, research, and academic trash such as paper, plastic, and glass.	hydrogen-3 halflife:12.32 y cobalt-60 half-life: 5.27 y
Class A:	Lowest level of LLRW, generally decays in 100 y.		
Class B:	Moderate level of LLRW, generally decay in 300 y.		
Class C:	Special controls required for this high level of LLRW, including shielding/barriers that must isolate. for 500 y.		
Greater than Class C:	Exceed the Class C limits and cannot be disposed in LLRW facilities; must be disposed with high-level radioactive waste.		
Exempt Material or Very Low Activity Waste	Material that is deemed a waste that contains trace concentrations of short half-life radionuclides that are considered below regulatory concern.	Various medical Procedures.	iodine-131 half-life: 8.027 days

Table: Typical characteristics of different waste classes.

Waste Classes	Typical Characteristics	Disposal Option
1. Exempt Waste (EW)	Activity levels or below clearance levels, which are based on a annual dose to members of the public less than 0.01m Sv.	No radiological restrictions.
2. Low and intermediate level wastes (LILW)	Activity levels above clearance levels and thermal power below about 2KW/m³.	
Short-lived waste (LILW-SL)	Restricted long lived concentrations (limitation of long lived alpha emitting radionuclides to 4000Bq/g in individual waste packages and to an overall average of 400 Bq/g per waste package).	Near surface or geological disposal facility.

Long lived waste (LILW-LL)	Long lived radionuclides concentrations exceeding limitations for short lived waste	Geological disposal facility.
3. High level waste (HLW)	Thermal power above about 2 KW/m^3 and long lived radionuclide concentrations exceeding limitation for short lived waste.	Geological disposal facility.

Management of Radioactive Wastes

Radioactive wastes contaminate air, water, soil and consequently vegetation that eventually affect public health and environment adversely. Since radioactivity is a long term phenomenon, proper and planned disposal of radioactive wastes should be ensured to prevent any negative impact on health and safety of living beings as well as on environmental quality, the principles of radioactive waste management are: (1) protection of human health, (2) protection of the environment, (3) protection beyond national borders, (4) protection of future generations and avoid burden over them (5) national legal framework, (6) control of radioactive waste generation, (7) management interdependencies, and (8) safety of facilities.

The major aim of the environmentalist is to abate the introduction of radioactive materials into the biosphere and environment. Control of the possible direct impact on the human environment is essential but not enough, since radionuclides can be transmitted through water, air, and land pathways for several years and even for generations. Some radioactive waste may be recovered and recycled by future reprocessing, but reprocessing process creates its own radioactive waste stream. The majority of radioactive waste can be treated only by isolating it from the available environment until its radioactivity no longer poses a threat. Isolation requirements vary for different classes of radioactive waste.

Table: Different kind of radioactive waste and its handling.

Type	Origin	Quantity (For 1000 MW station)	Disposal
Low level (LLW)	Waste produced during reactor maintenance	50 m3 /yr (Including packaging)	Packed in drum and shallow burial
Intermediate level (ILW)	Waste produced during reactor operation (reactor coolant purification)	50 m3 /yr (Including packaging)	Packed in drum and shallow burial
High level (HLW) R	Radioactive products extracted from spent nuclear fuel (SNF)	3 m3 /yr (Including packaging)	Vitrified packed in capsule and underground repository

A few radionuclides, mainly those that make up HLW, contain half-lives of tens of thousands of years, or even hundreds of thousands of years. It is difficult to find the technology that solves the problem of ultimate disposal of radioactive waste so long term storage is the only solution till date.

Segregation of Radioactive Waste

The secure, environmentally acceptable and economic management of radioactive waste arising from the nuclear power is an essential component of nuclear power programmes. Noteworthy technical advancement has been achieved and substantial experience of all stages of waste management has been accumulated in many countries. Waste processing, conditioning, storage and transportation have become regular and demonstrably safe activities.

Radioactive wastes should be separated according to their isotopes and placed in separate containers. Tritium (3H) and Carbon-14 (14C) can occupy the same container. Radioactive waste besides segregation based on isotopes, are also separated by physical form. Ten basic physical forms of radioactive waste include:

- Solid
- Glass
- Sharps
- Liquid
- Liquid Scintillation Vials
- Biological
- Animal Remains
- Source Vials
- Lead Pigs
- Sealed Sources

Radioactive Solids

It consists of disposable items, radioactive material contaminated lab items such as plastic/rubber gloves, tubing, and syringes; unbroken glassware such as pipettes, beakers, flasks, columns, etc. Specific type of containers is used for collection and storage of solid radioactive waste. Plastic bags are commonly used to collect the solid radioactive waste materials. Yellow radioactive material bags should be avoided for this type of waste expect when they are placed in closable waste receptacle (for e.g. plastic foot operated trashcan) which remains closed always. The prerequisites for safe handling and disposal of solid radioactive waste include:

- Extreme care while handling radioactive wastes.
- Proper labeling of all bags with the date, month, year, isotope, total activity and department's name.
- Before placing waste into the bags, remove all radioactive and labware labels.
- All the bags should be checked for leaks prior to waste collection. Sometimes, second yellow bag can be used to contain the waste if necessary.
- Take care to segregate sharps to avoid tearing of bags.
- Avoid placing of bags with solid radioactive waste in such places where there are chances of the waste be picked up by housekeeping personnel and disposed off along with municipal waste.
- Waste containers must always remain closed.

Radioactive Glass

It included radioactive material contaminated glassware/other unbroken glass which should be packaged separately from other solid radioactive waste. A strong cardboard box properly labeled and sealed is used for storing the radioactive glass.

Radioactive Sharps

Sharps are defined as anything that could tear the yellow radioactive material bag. Examples: Needles, broken glass, glass pipettes, razor blades, capillary tubes, etc. The container used for collecting such waste must be puncture resistant plastic tubes. These containers once full must be securely capped with orange/red top with complete labelling. The requirements for safe handling and disposal of this type of radioactive waste include:

- Care must be taken to avoid injury during packing of sharps in containers.
- Avoid overfilling of container.
- Ensure that all sharps are dry before placing into container.
- When full, securely cap tube with orange/red top.

Radioactive Liquid

The radioactive liquid waste can be further divided into:

- Aqueous: Aquous liquids are those water based based liquids with a pH between 5.0-9.0, such as saline and buffer solutions that arise from contaminated laboratory glassware washings; weak acids or bases free of biological, pathogenic, or infectious materials.
- Organic: laboratory solvents such as alcohols, aldehydes, ketones, and organic acids used for various purpose.
- Other liquids: Example contaminated pump oil.

Specially designated containers termed as carboys are used for disposing radioactive liquids. These carboys should be filled upto 80% of its volume. The prerequisites for safe handling and disposal of this type of radioactive waste include:

- In the laboratory the carboys should be provided with double containment as a precautionary measure against leakage. This will control carboy failure, necessary because pouring is usually accompanied by drips, dribbles, and seeping. Carboys should be placed in a tray or pan that will contain the liquid in the event of a spill of the carboy. At a minimum, plastic backed absorbent paper shall be placed under all liquid waste carboys.
- Do not use glass containers for storage of radioactive liquid waste. If plastic-incompatible contaminated organic solvents are required to be kept in glass containers, the bottle must be doubled contained.
- The first three rinses of the emptied containers must also be placed in the radioactive liquid waste container. No radioactive liquid is to be poured down the sink.

- Avoid mixing of liquid waste types in the carboys (e.g. organic with aqueous).
- Pipettes and other such items should not be placed in the carboys.
- All biological material in the carboys must be properly deactivated. Carboys should be kept free of contamination as far as possible.

Radioactive Liquid Scintillation Vials

Glass or plastic vials containing organic or aqueous based liquid scintillation fluid are included under this category. This type of waste is disposed of in the cardboard trays and placed in a yellow radioactive material bag. At certain cases, these wastes are double bagged in a yellow radioactive material bag for disposal Eg. glass vials are double bagged in yellow radioactive material bags and placed in a cardboard box. Absorbent material must be placed in the cardboard box to absorb any leakage from the vials. Empty vials are to be disposed of as dry and semi-solid radioactive waste. The requirements for safe handling and disposal of this type of radioactive waste are:

- Assure that all tops are fixed tightly on the vials.
- Avoid empting used vials into plastic bags. Leaking or seeping scintillation fluid will dissolve plastic.
- Label each box of vials containing aqueous scintillation cocktail and segregate them from organic cocktail vials.
- If the vials require reusability then pour the liquid waste into a carboy along with the first rinse of the vial. Because most scintillation cocktail solutions are volatile and will contain radioactive substances. The procedures of pouring out the solution should be conducted in a ventilated fume hood.

Radioactive Biological Waste

Radioactive waste containing biological, pathogenic, or infectious material and the equipment used to handle such material are categorized under radioactive biological waste. Examples: By-product animal waste (i.e. serum, blood, excreta), contaminated capillary tubes and other equipment contaminated with animal fluids, radioactive material labeled culture media. Yellow radioactive materials bags labeled with biological waste stickers or red biological bags labeled with radioactive material stickers are the container used for disposal. The prerequisites for safe handling and disposal of this type of radioactive waste include:

- Liquids must be absorbed into another material such as paper towels, sponges, gauze, etc. prior to placing into bags.
- Pathogenic and infectious waste must be sterilized by autoclaving or chemical treatment. If autoclaved, the autoclave must be checked for radioactive contamination after use.

Radioactive Animal Remains

Examples: Radioactive animal carcasses, animal bedding, and by-product animal waste with the carcasses (i.e. viscera, serum, blood, excreta, tissue, etc.) and other animal tissue containing radioactive materials. Yellow Radioactive Materials bags are used as containers for disposing these wastes. The requirements for safe handling and disposal of this type of radioactive waste are:

- Animal remains containing radioactive material in any quantity are subject to handling according to the guidelines. Avoid dumping of these wastes along with municipal wastes.
- A tag showing: the date, radioisotope, total activity, and the laboratory shall be tied to the bag. Bags not labeled cannot legally be disposed of, so an investigation will be performed to identify the generator of the unlabeled waste. If the bag is placed in the freezer by personnel from your lab, the same information must appear on the tag.
- Liquids surrounding carcasses must be absorbed onto another material (e.g. paper towels, sponges, gauze, etc.) prior to placing into yellow bags.

Source Vials Examples

The original vials with radioactive materials shipped from the supplier are categorized under this waste. This includes full, partially full, and empty vials. Source vials must be separated from the dry and semi-solid waste stream and placed in cardboard box for disposal.

Lead Pigs

Examples: Original lead and lead impregnated shielding containers surrounding source vials are included in this category. Lead is a hazardous waste and must be disposed of accordingly. Avoid placing of these items in municipal waste. Cardboard shipping box for used disposal.

All Sealed Sources

Examples: Calibration sources, check sources, quenched standard sets, electron capture gas chromatograph detectors, etc. Sealed sources must be separated from the solid waste and placed in cardboard box for disposal. During disposal, check for broken or crushed sources and handle these damaged sources with extreme care. All sources must be disposed of properly by a private contractor, even if decayed. A final survey or leak test on all sources should be done prior disposal.

Disposal of Radioactive Waste

Nuclear waste from commercial reactors, medical applications, defense industry, radioactive research poses a great threat to the scientific community as well as the public. The solutions to high-level nuclear waste are still debatable, both technically and ethically. There are numerous proposals for disposing high-level nuclear wastes. However, the most favored solution for the discarding of radioactive wastes is segregating out radioactive waste from man and biosphere for a long period of time such that any prospective release of radionuclides from the waste repository will not occur thereby resulting in unwarranted radiation exposure. The primary idea behind this solution is to use stable geological environments that can held the radioactive material for millions of years to provide an appropriate isolation capacity for the duration required.

The main reason for relying on such geological environments is based on the following chief consideration: 'Geological media is an entirely passive disposal system with no requirement for continuing human involvement for its safety. It can be abandoned after closure with no need for continuous surveillance or monitoring. The safety of the system is based on multiple barriers, both engineered and natural. The main one being the geological barrier.

Numerous alternatives for the disposal of long-lasting wastes have been examined over the years. Studies have been performed on concepts such as disposal in oceanic sub-seabed sediments and ultradeep boreholes, besides on more exotic proposals such as disposal into geologic sub-duction zones, in polar icecaps and launching into space. All of these concepts have been found inadequate in terms of costs or risks, or unfeasible owing to political or legal restrictions.

Most important problems due to legal, social, political and financial reasons have arisen in implementation due to:

- Environmental perceptions.
- Lack of proper awareness and education.
- 'Not-in-my-backyard' syndrome.
- 'Not-in-the-ocean' syndrome.
- Lack of proven technology.

Geologic Disposal

For handling long-lived radioactive waste, disposal in deep geological formations: under continental crust or under seabed has been recognized since 1957. Deep geological disposal involves introducing waste packages in underground structures dug in an impermeable geological medium having favourable properties in terms of its geological stability, hydrogeology, geochemistry and response to mechanical and thermal stress. The chosen medium must avoid areas of exceptional interest in terms of exploitable underground resources. The structures must be located at least 200 m below the ground surface to avoid the effects of erosion and human intrusion. Depositing the waste in deep geological sites will favor isolation of waste and keep them stable over thousands of years. Long-lived radioactive wastes are buried in deep sites unlike low-level radioactive waste which is normally disposed in near-surface facilities or abandoned mines.

Mostly, crystalline (granitic, gneiss) or argillaceous (clays) or salty or tuff are preferred for disposal of high-level radioactive waste. Provisional storage facilities, which permit cooling of the wastes over a few decades also do exist in most of the countries. The majority of countries now believe deep geological disposal as the standard solution for final management of high-level and intermediate-level long lived waste. Granite; sedimentary formations, more especially clay beds; and salt are three types of geological formations on which most of the in progress research and studies are mainly focusing.

Underwater Disposal

Ocean-Dumping

Generally, two associated methods of underwater disposal of Spent Nuclear Fuel (SNF) exist: dumping containers of radioactive waste into the ocean, and sub-seabed disposal. The main reason of underwater disposal of SNF is the same as any other type of SNF disposal, which is to isolate radioactive waste from human contact and the environment long enough for any release of radiation to become harmless.

USA, France, Great Britain, and many more industrialized countries have opted dumping into the oceans as the least expensive method for disposal of the radioactive wastes. Before the U.S. government banned or stopped sea dumping of spent nuclear fuel (SNF) around the 1980s, the United States reportedly dumped about 112,000 containers of nuclear waste at thirty locations in the Atlantic and Pacific oceans.

Although this practice has been prohibited by nearly all of the countries with nuclear programmes, the problem still persists. Russia, which presently controls sixty per cent of the world's nuclear reactors, continues to dispose of its nuclear wastes into the oceans. According to Russia's Minister of Ecology, it will continue to dump its wastes into the oceans because it has no other alternative method. It will continue to do so until it receives enough international aid to create proper storage facilities. In response, the United States has pledged money to help Russia, but the problem continues. Although radioactive waste has known negative effects on humans and other animals, no substantial scientific proof of bad effects on the ocean and marine life has been found. Therefore, some nations have requested that ocean-dumping should be carried on. Others argue that the practice should be banned until further proof of no harm is available.

Sub-Seabed Disposal

Seabed disposal involves disposal of waste in deep oceans floor situated 5 km below the sea surface. These large tectonic plates are desert-like and are covered with hundreds of meters of thick sedimentary soft clay. Sub-seabed disposal is complexed and involves two associated methods: creating a repository with a possibility for retrieval of SNF, and everlasting burial.

Sweden and the United Kingdom consider creating a repository method, which permits for retrieval of SNF and might, contains admittance to the repository from land. A benefit of such a sub-seabed repository would be an increased capability to supervise SNF, in contrast to the dumping method. Another advantage could be a potential for access from land, which could let the repository method to avoid violating international bans against oceanic dumping, but accessing certain underwater locations by land might be impossible.

In Seabed Burial method, the mud flats at the bottom of the ocean is drilled to depths of the order of hundreds of meters. Inside these bore holes the radioactive material that are enclosed in containers are stacked vertically one over the other. Inconsistency of the rock and high local permeability limits the use of rocks in ocean as basement. Moreover, oceanic water has a mixing time of the order of a few thousand years which does not serve as a good barrier for long-lived radionuclides.

Experimental works have already established that clays have the property of holding on to several radioactive elements, including plutonium; hence, seepage of these elements into saline water is negligible. Rates of migration of these elements over hundreds of thousands of years would be of the order of a few meters. Hence, during such long times, radioactivity will lessen to levels below the natural radioactivity in sea water as a result of natural radioactive decay. The clays also contain plastic-like behaviour to form natural sealing agents. Lastly, the mud-flats have rather low permeability to water; hence, leaching probability is rather low. It may be noted that the method depends on standard deep-sea drilling techniques routinely practiced and sealing of the bore-holes. Core samples from about half a dozen vastly separated sites in the Pacific and Atlantic oceans have 'showed a continuous history of geological stillness over the past 50–100 million years'. However, there are several questions that still remain to be answered:

- Whether migration of radioactive elements through the ocean floor is at the same rate as that measured in the laboratories?
- What is the effect of nuclear heat on the deep oceanic-clays?
- What is the import on the deep oceanic fauna and waters above?
- In case the waste reaches the seabed-surface, will the soluble species (for example, Cs, Tc, etc.) be diluted to natural background levels? If so, at what rate?
- What occurs to insoluble species like plutonium?
- What is the possibility of radioactivity reaching all the way to the sea surface?
- In problems of accidents in the process of seabed burial leading to, say, sinking ships, to loss of canisters, etc. how does one recover the waste-load under such scenarios?
- What is the likelihood that the waste is hijacked from its buried location?

Other problems include:

- International agreement to consider seabed-burial as distinct from 'ocean-dumping'.
- This method would be expensive to implement, but its cost would be an impediment to any future plutonium-mining endeavor.

Underwater disposal of SNF, specifically sub seabed disposal, can occur far from the coast of any state or nation and may perhaps thereby avoid the NIMBY ("not in my backyard") syndrome, but this result is not guaranteed considering existing laws and a popular belief that Earth's oceans are a global common.

Though the world tendency is in the direction of the option of land-based disposal, it is unsure whether confining repositories to land-based sites really assist in prevention of sea pollution. If radionuclides from a land-based repository leached out to the surface, they would be rapidly transported to the sea by surface water. The most essential thing is to isolate radionuclides from the biosphere as consistently as possible. If sub-seabed disposal results in more dependable isolation, sub-seabed disposal is the better safeguard against sea pollution. This method takes into consideration technological feasibility, protection of marine environments, and availability of international understanding. The United Nation's Convention on the Law of the Sea defines that a coastal state is granted sovereign rights to utilize all resources in water and under the seabed within its exclusive economic zone (EEZ), which can extend from the coast line up to 200 nautical miles (about 370 km) offshore. A repository is proposed to be constructed in bedrock 2 km beneath the seabed. To utilize sub-seabed disposal within the EEZ, it is also proposed that waste packages would be transported through a submarine tunnel connecting land with the sub-seabed repository. Sea pollution by an accident during disposal work would be unlikely, because waste would never go through sea water during the work. The proposed method is a deviation of geologic disposal. Long-term monitoring is also possible by maintaining the access tunnel for some time after constructing artificial barriers. Everlasting burial is an additional method of sub-seabed disposal and might include at least three options: (1) Shallow penetration of the seabed, (2) deep burial through drilling, and (3) subduction involving tectonic dissipation of SNF. These options are not strictly separated categories. They are linked approaches that share alike features and could result in similar outcomes.

Subductive Waste Disposal Method

It is the latest disposal technology for nuclear wastes. This method is considered important because the material deposited will not return to the biosphere. It eliminates the material completely from the biosphere faster than it can return. Subductive method is safest, most sensible, most economical, most stable (for long term) and environmentally benign. It is considered safe because during subduction, the waste will be carried to the interior of the earth, diluted and dispersed. Subduction is a process whereby one tectonic plate slides beneath another and is ultimately reabsorbed into the mantle. The rate at which the plates subduct is 2.1 cm per annum which is faster than diffusion of radionuclides (1 mm annually) through the turbidite sediments. The movement of these plates is very slow, during which vertical fractures over a repository at the interface are sealed.

Transmutation of High-Level Radioactive Waste

This means of high-level radioactive waste foresees that one may use transmutational devices, consisting of a hybrid of a subcritical nuclear reactor and an accelerator of charged particles to 'destroy' radioactivity by neutrons. 'Destroy' may not be the proper word; what is effected is that the fission fragments can be transmuted by neutron capture and beta decay, to produce stable nuclides. Transmutation of actinides involves several competing processes, namely neutron-induced fission, neutron capture and radioactive decay. The large numbers of neutrons produced in the spallation reaction by the accelerator are used for 'destroying' the radioactive material kept in the subcritical reactor. The scheme has not so far been demonstrated to be practical and cost effective.

Solar Option

It is proposed that 'surplus weapons' plutonium and other highly concentrated waste might be placed in the Earth orbit and then accelerated so that waste would drop into the Sun. Although theoretically possible, it involves vast technical development and extremely high cost compared to other means of waste disposal. Robust containment would be required to ensure that no waste would be released in the event of failure of the 'space transport system'.

There has been by a large agreement amongst the waste management community that deep geologic disposal represents the only possible route for ensuring sufficient long-term isolation of long-lived waste from the human environment and thus "closing the nuclear fuel cycle". Current debates within waste management programmes focus mostly on when and where disposal should take place, bearing in mind the need to fulfill ethical obligations, reduce present and future risks, ensure that other management options are given due consideration, and enhance societal participation in and ultimately acceptance of the disposal strategy.

Effects of Radioactive Waste

Everything on Earth is exposed to radiation. However, exposure to radiation at levels greater than natural background radiation can be hazardous. If radioactive waste is improperly disposed of or if the disposal methods are compromised, there can be serious consequences and effects. High-level radioactive waste from nuclear reactors can be hazardous for thousands of years.

Health Effects

Exposure to certain high levels of radiation can have negative effects on the human body that can even cause death. Radiation exposure can also cause cancer, birth defects, and other abnormalities, depending on the time of exposure, amount of radiation, and the decay mechanism. It is interesting to know that we are exposed to radiations naturally by living the lives that come from the ground below us. Radiation can cause changes in 'DNA' that ensures cell repair.

Scavenging

Scavenging of abandoned radioactive material has been the cause of several other cases of radiation exposure, mostly in developing nations, which may have less regulation of dangerous substances and a market for scavenged goods and scrap metal. Irresponsibility on the part of the radioactive material's owners, usually a hospital, university or military, and the absence of regulation concerning radioactive waste, or a lack of enforcement of such regulations, has been significant factors in radiation exposures.

The scavengers and those who buy the material are almost always unaware that the material is radioactive, and it is selected for its aesthetics or scrap value. Usually, people who scavenge these sorts of materials will end up in the hospital and may even die of problems related to or caused by the radioactive materials.

Accidents

When radioactive material was disposed of improperly or safeguarding during transport was defective, or when it was simply abandoned or even stolen from a waste store, a few incidents have occurred. In the Soviet Union, waste stored in Lake Karachay was blown over the area during a dust storm after the lake had partly dried out.

At Maxey Flat, a low-level radioactive waste facility located in Kentucky, containment trenches covered with dirt, instead of steel or cement, collapsed under heavy rainfall into the trenches and filled with water. The water that invaded the trenches became radioactive and had to be disposed of at the Maxey Flat facility itself.

In other cases of radioactive waste accidents, lakes or ponds with radioactive waste accidentally overflowed into the rivers during exceptional storms. In Italy, several radioactive waste deposits let material flow into river water, thus contaminating water for domestic use.

Water Problem

If radioactive waste seeps into the ground, drinking water can become contaminated, too, which is absolutely disastrous for locals and residents live close to the epicenter of the disaster. It can eventually get into reservoirs and other water sources and, from there, can reach the homes of people who unwillingly drink high radioactive material.

Transportation Hazards

Despite all the cautions that are put into place while transporting radioactive waste from power plants, accidents still occur and can have a devastating effect on all those in the vicinity of the crash can occasionally result in problems. In case poor shipping casks are used for the containment of radioactive material, then a slight knock or bump or even crash could cause the contents to spill and affect a wide radius.

Expensive Cleaning

If one of these accidents does occur, there is no simple or easy route when trying to clean up spilled radioactive material. The cost of cleaning up and making everything safe once again for people, animals and plants is very high, and it can take years to ensure that an area is safe to live in or even to visit once again. In the case of very serious accidents, it may take many decades until things normalize to start living once again.

E-Waste

Electronic waste or 'e-waste' in general refers to electrical and electronic equipment (EEE) that has ceased to be of any value to its owners. E-waste is any electrical or electronic equipment that's been discarded. This includes working and broken items that are thrown in the garbage or donated to a charity reseller like Goodwill. Often, if the item goes unsold in the store, it will be thrown away. E-waste is particularly dangerous due to toxic chemicals that naturally leach from the metals inside when buried.

E-Waste Generation

The major problem associated with e-waste management is its ever increasing quantum. However, the e-waste quantities represent a small percentage of the overall municipal solid waste (MSW). Data on e-waste generation may vary between areas of a country because of the definitions of waste arising, technological equipment used, the consumption patterns of the consumers, and changes in the living standards across the globe. Global e-waste generated per year amounts to approximately 20-25 million tons, most of which is being produced in rich nations such as the United States (US) or European Union member countries. The US, is the largest generator of e-waste, with a total accumulation of 3 million tons per year; and China is the second largest, producing 2.3 million tons each year. Brazil generates the second greatest quantity of e-waste among emerging countries.

In Malaysia, the volume of e-waste generated is estimated at roughly 0.8-1.3 kg of waste per capita per day, with an increasing trend of e-waste generation, which rose to 134,000 tons in 2009. Furthermore, the volume of e-waste in Malaysia is expected to rise to 1.1 million metric tons in 2020, at an annual rate of 14%. In South Africa and China, e-waste production from old computers will increase by 200-400% from 2007 to 2020, and by 500% in India. In this same period e-waste from televisions will be 1.5-2 times higher in China and India; whereas in India, e-waste from discarded refrigerators will double or triple by 2020. For India, the volume of e-waste generated is 146,000 tonnes per year. However, these data only include e-waste generated nationally and do not include waste imports (both legal and illegal) which are substantial in emerging economies such as India and China. The reason is that large amount of WEEE enters India from foreign countries without paying any duty in the name of charity. The rate at which the e-waste volume is increasing globally is 5 to 10% yearly.

Composition of E-Waste

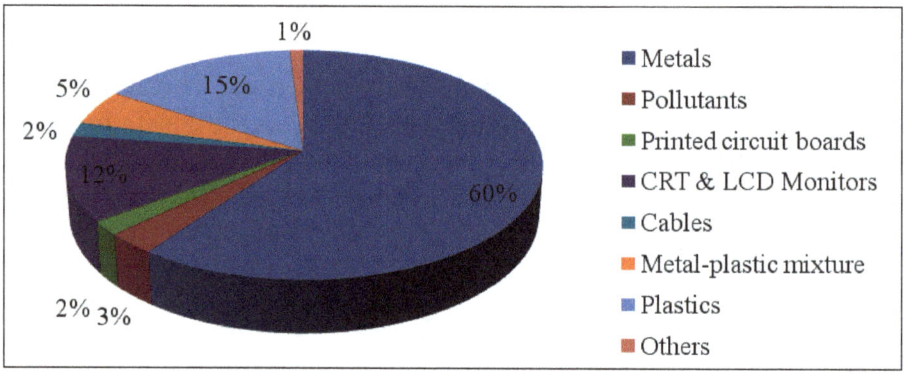

Figure: The distinctive contents of a WEEE.

E-waste normally contains valuable, as well as potentially toxic materials. The composition of e-waste depends strongly on factors such as the type of electronic device, the model, manufacturer, date of manufacture, and the age of the scrap. Scrap from IT and telecommunication systems contain a higher amount of precious metals than scrap from household appliances. For instance, a mobile phone contains more than 40 elements, base metals such as copper (Cu) and tin (Sn); special metals such as lithium (Li) cobalt (Co), indium (In), and antimony (Sb); and precious metals such as silver (Ag), gold (Au), and palladium (Pd). Special treatment of e-waste should be considered to prevent wasting valuable materials and rare elements. Materials such as gold and palladium can be mined more effectively from e-waste compared to mining from ore. By contrast, e-waste contains PBDEs, which are flame retardants that are mixed into plastics and other components. Circuit boards found in most of the electronic devices may contain arsenic (As), cadmium (Cd), chromium (Cr), lead (Pb), mercury (Hg), and other toxic chemicals. Typical printed circuit boards treated with lead solder in electronic devices contain approximately 50 g of tin-lead solder per square meter of circuit board. Obsolete refrigerators, freezers, and air conditioning units contain ozone depleting Chlorofluorocarbons (CFCs).

The prominent materials such as barium, cadmium, copper, lead, zinc, and other rare earth metals are contained in end-of-life (EOL) cathode ray tubes (CRTs) in computer monitors, and televisions. For example, items such as leaded glass provide protection against X-rays produced in the picture projection process in CRTs. The average lead in CTR monitors is 1.6-3.2 kg. Thus, the US and other developed countries in the EU and Japan have banned the disposal of cathode ray tubes in landfills because of their toxic characteristics. A critical challenge in designing and developing strategies to manage e-waste is the changing composition of the many constituents due the advancement of technology, particularly in the electronic components. It is against this background that e-waste recycling and disposal methods ought to keep pace with the changing composition of e-waste. Several factors influence the composition of e-waste, including economic conditions, availability of a reuse market, and infrastructure of the recycling industry, waste segregation programs, and regulation enforcement. Figure illustrates the distinctive materials in a WEEE.

E-Waste Data for Several Countries Across the Globe

Amount of E-Waste Collected and Treated

E-waste generated from the different diverse sources is normally collected as a whole unit or sub-unit of functional equipment. In many instances across the globe, whole units of e-waste have been categorized as e-waste. Based on the number of discarded information communication technology (ICT) devices collected in Europe, computers, cell phones, fixed-line telephones, televisions, and radios are the major electronic products, and together they amounted to 11.7 million tons in 2007. In 2004, approximately 75,000 tons of WEEE were collected, classified, disassembled, and then processed in Switzerland, compared with the collection of approximately 68,000 tons in 2003.

In developing and transition countries, little consideration is given to the quantification of the e-waste collected. The reason is that in pre-reprocessing stages, collection of the e-waste is mostly undertaken by the unorganized sector of scrap dealers/traders or peddlers. As a result, this information is invisible to the statistics collection system, which makes quantification of e-waste very difficult in developing and transition countries. More precise figures regarding unused electronic and electrical equipment/waste electronic and electrical equipment (UEEE/WEEE) are not available because the customs data do not distinguish between used and new equipment and the import statistics reveal only total values. Based on the current understanding on e-waste management, research studies suggest that to achieve sustainable development goals associated with waste management would require successful establishment of baseline levels of information from which more informed e-waste management and policy decisions can be made. Similarly, to effectively manage e-waste could require establishment of separate collection channels that would be environmentally friendly. This could result in the reduction of e-waste generated and its environmental impacts.

In the EU, the EU WEEE directive clearly imposes collection, recovery, and recycling targets on its member countries. Thus, it stipulates a minimum collection target of 4 kg/capita per year for all the member states. These collection- and weight-based recycling targets seek to reduce the amount of hazardous substances disposed into landfills and to increase the availability of recyclable materials that indirectly encourages less virgin materials consumption in new products. Switzerland is the first country in the world to have established and implemented a formal e-waste management system that has recycled 11 kg/capita of WEEE against the target of 4 kg/capita set by the EU. One-third of electrical and electronic waste in the EU is reported as separately collected and appropriately treated.

In 2006, Germany collected and treated about 754,000 tons of e-waste according to the ElektroG system, while other EU member states collected about 19,000 tons. It was also forecasted that IT and telecommunications equipment put on the market were 315,000 tons, and the waste collected and treated in the system according to ElektroG was about 102,000 tons (7,000 tons of this was collected from other EU countries). This shows the effective collection and treatment of e-waste in the EU. The introduction of the extended producer responsibility (EPR) scheme in 2003 was the most important step in South Korea, and about 70% of e-waste was collected by producers. Over the same period, the amount of e-waste reused and recycled was 12% and 69% respectively. The remainder was sent to landfill sites or incineration plants, accounting for 19%.

Amount of E-Waste Disposed

The scientific and environment friendly disposal of e-waste is critical. Relevant past studies on e-waste management confirmed that rapid growth combined with rapid product obsolescence are the most important factors making discarded e-products one the fastest growing waste fraction, accounting for 8% of all municipal waste in the EU. If not disposed of properly it could lead to significant negative environmental impacts. The average for developing and transition countries was 1% of total solid waste, which increased to 2% in 2010. Developing and transition countries, especially those in Africa and Asia, are the primary destinations for e-waste dumping, despite these countries lacking basic disposal technologies or facilities.

In 2012, more than 70% of the total electronic waste collected worldwide was actually exported or discarded by developed countries. In the US alone, 130,000 computers and more than 300,000 cell phones are disposed each day, and an estimated 80% of the generated e-waste is sent to less-developed countries. In 2007 in the US, 410 thousand tons were recycled (13.6%), and the rest was improperly discharged in landfills or incinerated. Between 2003 and 2005, approximately 80-85% of the e-waste ready for EOL management ended up in US landfills. A related study about e-waste management in the US pointed out that in 2009, enormous quantities of e-waste (82.3%) was disposed in landfill sites and incinerators, while 17.7% went to the recyclers. In the EU, it is shown that two-thirds of this waste stream is potentially still going to landfills and to sub-standard treatment sites in or outside the EU. In China, huge volumes of e-waste have been discarded in recent years as people more frequently replace their old home appliances with new ones.

A relevant case-study on e-waste management pointed out that it is not possible to make an overall comparison between different countries, even if they are in the same continent, as the definitions in legislation and categorization of e-waste streams differ. Nevertheless, it is established that the main volumes of e-waste reside in developed countries.

Collection, Treatment and Disposal Systems

Collection, treatment, and disposal systems are critical elements of e-waste management. Most developed countries have framed conventions, directives, and laws aimed at fostering proper collection, treatment, and recycling of e-waste, as well as safe disposal of the non-recyclable components. These include the EPR, product stewardship, advance recycling fund (ARF), the 3Rs or Reduce, Reuse, Recycle initiative, etc. For the EU, two directives have been promulgated to place an obligation on the producers of e-goods to take back EOL or waste products free of charge in an effort to reduce the amount of waste going to landfills. However, in developing and transition countries, e-waste is treated in backyard operations, using open sky incineration, cyanide leaching, and simple smelters to recover precious metals mainly copper, gold, and silver—with comparatively low yields—and discarding the rest with municipal solid waste at open dumps, into surface water bodies and at unlined and unmonitored landfills, thereby causing adverse environmental and health effects. Table presents a comparison of typical e-waste treatment processes in developed and developing countries.

Table: Comparison of typical e-waste treatment processes in developed and developing countries.

Developing countries	Developed countries
"Informal" sector	Formal sector
Manual dismantling	Manual dismantling
Manual separation	Semi-automation separation
Recovery of metals by heating, burning, and acid leaching of e-waste scrap in small workshops.	Recovery of metals by the state-of-the-art methods in smelter and refineries.

Disposal System

Disposal of e-waste is mainly through landfilling. Most often, the discarded electronic goods finally end-up in landfill sites along with other municipal waste or are openly burnt releasing toxic and carcinogenic substances into the atmosphere. In developing and transition countries the disposal of e-waste in the informal sector is very rudimentary so far as the safe techniques employed and practices are concerned, resulting in low recovery of materials. Table presents a comparison of typical disposal systems in developed and developing countries.

Table: Comparison of a typical e-waste disposal systems in developed and developing countries.

Developed countries	Developing countries
Incineration with MSW	Opening burning
Landfill disposal	Open dumping

E-waste management is different between developed countries and developing and transition countries. Developing and transition countries do not have guidelines and information campaigns on the fate of e-waste. Especially, less sophisticated disposal systems are used, from open burning and dumping to uncontrolled landfill sites, which pose significant environmental pollution and occupational exposure to e-waste-derived chemicals. Serious challenges in the disposal of e-waste were analyzed across developing countries such as Brazil, China, and India, outlining the difficulty to implement/enforce existing regulations and clean technologies backed by lack of capacity building and awareness. In contrast, developed countries have devised sophisticated disposal schemes and high-cost systems, which are less hazardous to handle waste. However, a comprehensive overview of the situation is constrained by the availability of data. This means that the differences in the socio-economic and legal contexts between typical developing and developed countries' scenarios limit e-waste management in developing and transition countries. The regulations that guide the disposition of e-waste in developing countries is mostly fragmented and lack monitoring, while in developed countries the regulations are stringent and there is effective monitoring.

Biomedical Waste

Biomedical waste, also known as infectious waste or medical waste is defined as waste generated during the diagnosis, testing, treatment, research or production of biological products for humans or animals. Biomedical waste includes syringes, live vaccines, laboratory samples, body parts, bodily fluids and waste, sharp needles, cultures and lancets.

Improper management of waste generated in health care facilities causes a direct health impact on the community, the health care workers and on the environment. The waste generated in these institutions essentially consists of solids and liquid, which may be hazardous, infectious and non-infectious. It has been estimated that up to 85% to 90% of the waste generated in hospitals is non-infectious (free with any body fluids, which is similar to domestic waste). It is the remaining 10% to 20% of waste that is of concern because it is hazardous and infectious. In addition, waste that is un-segregated and not treated in the right manner would cause environmental pollution affecting the health of the community. From waste audits done at several hospitals by a few NGOs, arrived at some figures, which can now be used and extrapolated for the whole country. These audits must be conducted only after adequate training on waste segregation is given to health care institutions.

Proper handling, treatment and disposal of biomedical wastes are important elements of health care office infection control programme. Correct procedure will help protect health care workers, patients and the local community. If properly designed and applied, waste management can be a relatively effective and an efficient compliance-related practice.

Proper collection and segregation of biomedical waste are important. At the same time, the quantity of waste generated is equally important. A lesser amount of biomedical waste means a lesser burden on waste disposal work, cost saving and a more efficient waste disposal system. Hence, health care providers should always try to reduce the waste generation in day-to-day work in the clinic or at the hospital.

Sources

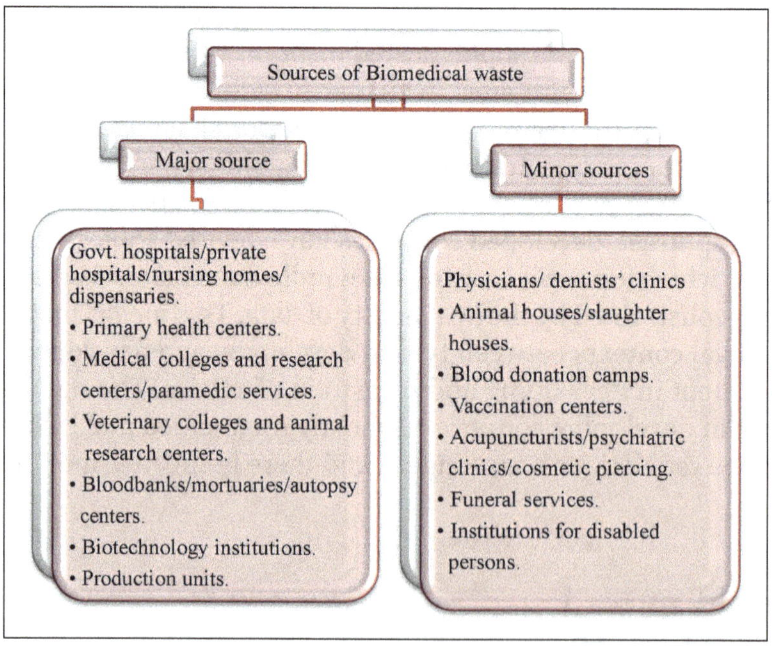

Figure: Sources of biomedical waste.

The source of biomedical waste is the place or the location at which biomedical waste has been generated. The source of biomedical waste is classified into two types based on the quantity of waste generated. They include major and minor source. Major source generates more amount of biomedical waste compared to minor source and also there is regular generation of biomedical waste in the major source which includes government hospitals, private hospitals, nursing home and dispensaries. Minor source includes physicians and dental clinics. Figure below shows the details of the various source of biomedical waste generation.

Classification

The classification of the biomedical waste is carried out based on its characteristics, source of generation and the level of hazard to the environment. The biomedical waste is classified into two types:

- Non-hazardous waste,
- Hazardous waste.

Non-Hazardous Waste

About 75% to 90 % of biomedical waste characteristics were similar to that of domestic waste and are non-risky in nature. This waste is generated mainly from the organization and maintenance of hospital and health care centers.

Hazardous Waste

The remaining 10 – 25% of biomedical waste falls under the hazardous waste categories. The hazardous waste contains infectious characteristics of about 15% - 18 % and toxicity characteristics of about 5% - 7%. The various hazardous wastes includes:

- Infectious waste: Waste containing pathogens; e.g. excreta; laboratory cultures; isolation wards waste; swabs, materials, or equipments that have been in contact with infected patients.
- Pathological waste: Human tissues or fluids e.g. body parts; blood and other body fluids; fetuses.
- Pharmaceutical waste: Waste containing pharmaceuticals; e.g. pharmaceuticals that are expired or no longer needed; contaminated pharmaceuticals (bottles, boxes).
- Genotoxic waste: Waste containing cytostatic drugs (often used in cancer therapy)/genotoxic chemicals.
- Chemical waste: Waste containing chemical substances e.g. laboratory reagents; film developer; disinfectants and solvents that are expired or no longer needed.
- Wastes with high content of heavy metals: Batteries, Broken thermometers, blood pressure gauges, Pressurized containers, gas cylinders, gas cartridges, aerosol cans.
- Radioactive waste from radiotherapy: Waste containing radioactive substances e.g. unused liquids from laboratory research; contaminated glassware, packages or absorbent paper; urine and excreta from patients treated or tested with uncapped radionuclide.

Biomedical Waste Management

Proper management of biomedical waste is highly essential since it induces various risk to the human health and to the surrounding ecosystem that leads to the ecological hazard, professional hazard and public hazard. Steps involved in biomedical waste management were shown in Figure.

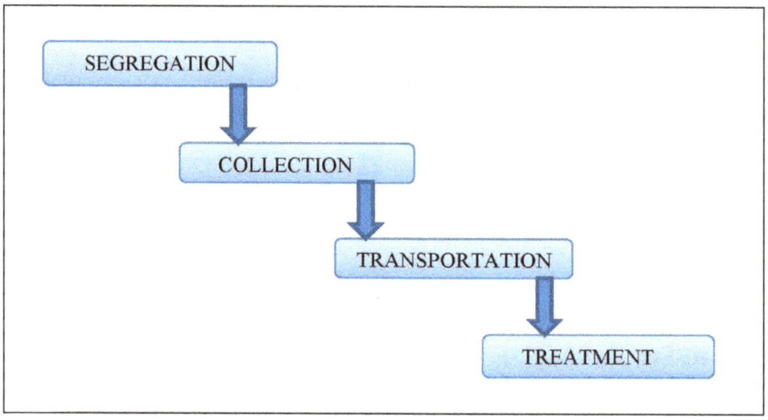

Segregation

To avoid mixing of the biomedical waste with other, a container should be set to the side with colour coding bags at the point of generation. The sorting or separation of waste into different categories is referred as segregation. Segregation will decrease or minimize the risks in addition to rate of managing and disposal. Segregation is the most important and critical step in bio-medical waste management. Only, effective segregation can confirm the effective bio-medical waste management.

How does Segregation Help?

Segregation plays an effective role in handling and treatment of waste. It reduces the quantity of waste and if done effectively, it can avoid the mixing of biomedical waste with any other type of waste especially municipal waste. Segregation will avoid the reuse of certain biomedical waste like used syringes, needles and other plastics. Some materials like plastics can be recycled after proper disinfection and these can be reused for non-food grade products.

During segregation process, the biomedical waste must be separated under the following categories shown in Table below. Category no.1 includes the Human anatomical waste in which the human tissues, organs, body parts are considered. Animal waste falls under the Category No. 2. It includes Animal tissues, organs, body parts, carcasses, bleeding parts, fluid, blood and experimental animals used in research, waste generated by veterinary hospitals and colleges, discharges from hospitals, animal houses. Category No. 3 is the Microbiology & Biotechnology waste which contain Wastes from laboratory cultures, stocks or specimen of live microorganisms or attenuated vaccines; human and animal cell cultures used in research; infectious agents from research and industrial laboratories; wastes from production of biologicals, toxins and devices used for transfer of cultures. The Category No. 4 includes waste Sharps in which Needles, syringes, scalpels, blades, glass, etc. that may cause puncture and cuts. This includes both used and unused sharps. Discarded Medicine and Cytotoxic drugs falls under the Category No 5 which consists of wastes comprising of outdated, contaminated and discarded medicines. The soiled waste is included in the Category No. 6 containing items contaminated with body fluids including cotton, dressings, soiled plaster casts, lines, bedding and other materials contaminated with blood.

Table: Categories of Waste.

Waste Category	Type of Waste
Category No. 1	Human Anatomical Waste
Category No. 2	Animal Waste
Category No. 3	Microbiology and Biotechnology Waste
Category No. 4	Waste Sharps
Category No. 5	Discarded Medicine and Cytotoxic drugs
Category No. 6	Soiled Waste
Category No. 7	Solid Waste
Category No. 8	Liquid Waste
Category No. 9	Incineration Ash
Category No. 10	Chemical Waste

Category No 7 is the solid waste which includes waste generated from disposable items other than the waste sharps such as tubing, catheters, intravenous sets, etc. Liquid waste falls under the category no. 8, it consists of waste generated from the laboratory and washing, cleaning, housekeeping and disinfecting activities. Category No 9 includes incineration ash i.e., ash from incineration of any biomedical waste. Chemical Waste falls under Category No 10 and consists of Chemicals used in production of biologicals, chemicals used in disinfection and as insecticides etc.

Collection and Storage

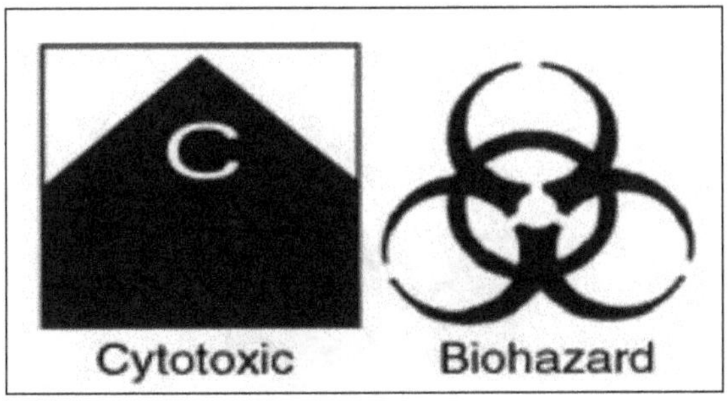

Figure: Symbols

The collection of biomedical waste involves the installation of different colour coded containers for biomedical wastes obtained from varying sources. The containers/ bins should be placed in a location so that 100 % collection is achieved. The bins and bags that hold the biohazard symbol as shown in figure represents the nature of waste. The symbols in biomedical waste management is generally used as a warning to take precautions while exposing to those substances. The biohazard symbol was developed by the Dow Chemical Company in 1966 for their containment products.

Subsequent to collection, the biomedical waste is stored in specific containers and stored in a proper place. The extent of storage should not exceed beyond 8-10 h in big hospitals containing more than 250 bedded and 24 h in nursing homes. Each container must be clearly labelled with the location being mentioned in them. The purpose of labelling is to trace the waste at the source. Storage spot must be clear with a warning sign.

Collection of the biomedical waste was carried out in its specific coloured bags. In the yellow colour bags, the categories 1,2,3 and 6 waste will be collected and this bags are made up of plastic materials. The Red bags are made up of disinfected container or plastic in which Category 3, 6 & 7 waste will be collected. The Blue/ White Translucent bags collect Category 4 & 7 waste which is made up of Plastic/ puncture proof container. The black coloured plastic bags are used for the collection of waste under category 5, 9 & 10. Figure below shows the collection of biomedical waste in the colour coded boxes.

Transportation

The collected wastes are transported in trolleys or in enclosed wheelbarrow for treatment. The operator should ensure to avoid manual loading. The bags / Container containing biomedical wastes must be tied/ lidded before hauling for treatment. Vehicles used for transporting should be special to avoid contact to, and direct contact with the operator, scavengers and the public. While transporting the containers, it must be properly enclosed. The effects of traffic accidents should be incorporated in the design, and the driver must be trained in the actions which must be followed in case of an accidental spillage. The interior of the containers should also be rinsed thoroughly.

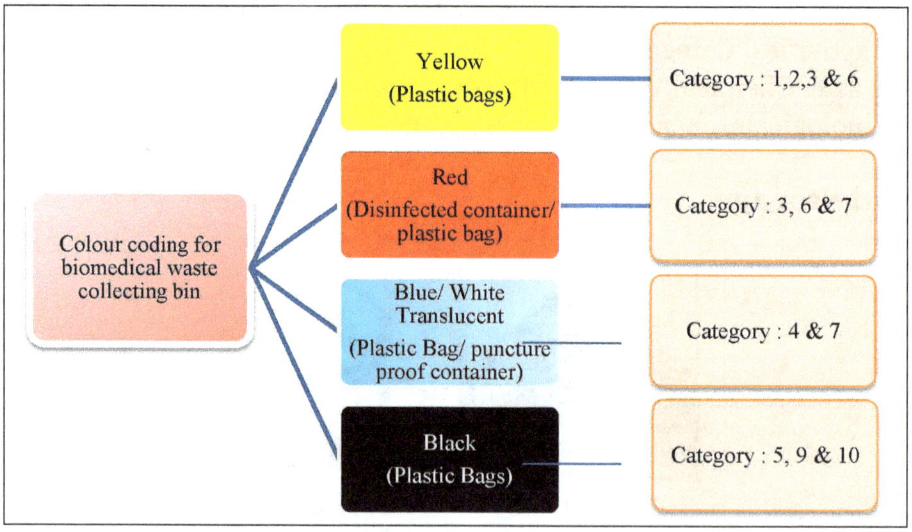

Figure: Collection of biomedical waste in a colour coded boxes.

Trolleys

The use of trolleys will make the elimination of infectious waste possible at the source itself, instead of accumulation a new category of waste.

Wheelbarrows

Wheelbarrows are used to transfer the waste from the point source to the collection centres. There are two types of wheelbarrow – covered and open. Wheelbarrows are made of steel and provided with two wheels and a handle. Open dumping should not be done. Only packed waste (in plastic bags) should be carried. To prevent corrosion, care should be taken to prevent the liquid waste from spilling into the wheelbarrow. Wheelbarrows also come in various sizes depending on the utility.

Chutes

Chutes are vertical conduits provided for easy transportation of biomedical waste vertically in case of establishment with more than two floors. Chutes should be produced from stainless steel. It should have a self-closing lid. These chutes have to be sterilized on a daily basis with formaldehyde vapours. The linen that are contaminated with blood or other body fluids from each floor must be bundled in soiled linen or in plastic bags before expelling into the chute.

Alternately, elevators with mechanical winches or electrical winches can be used to bring down waste containers from each floor. Chutes are essential to keep away from horizontal transport of waste thereby diminishing the routing of the waste within the premises and hence reducing the risk of secondary contamination.

Dustbins

It is very important to calculate the amount of waste generated at each point. Dustbins should be of such capacity so that it can be placed at this specific site and that they do not overflow between each cycle of waste collection. Dustbins have to be cleaned subsequently at each cycle of clearance of waste with disinfectants. Dustbins can be wrinkled with plastic bags, which are chlorine-free, and colour coded as per the law.

Treatment and Disposal

Before its final disposal of biomedical waste, it must be disinfected. Anatomical waste can be disposed by deep burial. Syringes to be cut (with hub cutters) and chemically disinfected with1% bleaching powder solution at source of generation before final disposal into sharps pit. Infected plastics to be chemically disinfected or autoclaved, shredded and recycled and sent for final disposal into municipal dumps.

Incineration

Most of the hazardous biomedical wastes was treated by the method of incineration to reduce organic and combustible waste to inorganic incombustible matter. Incineration is a high temperature, dry oxidation process that results in significant reduction of waste volume and weight. Wastes that cannot be reused, recycled or pose problem in disposing in landfills are treated by incineration. Examples of wastes that cannot be incinerated are chemical wastes, wastes containing high mercury or cadmium (broken thermometers, second-hand batteries, and lead lined wooden panels, sealed ampules or ampules containing heavy metals), silver salts, pressurized gas containers, photographic or radiographic wastes, halogenated plastics such as PVC.

The advantages of incinerator include high reduction of waste volume in addition to good disinfection competence. It helps to save the space in the landfill. The ash generated can be disposed of safely in the landfills. The major disadvantage of incineration includes high operating cost as they are energy intensive process. Also it releases a huge amount of atmospheric pollutants. The need for cyclic removal of slag and dirt, inadequacy in demolishing anti-thermal chemicals and drugs such as cyto toxic are its other disadvantages.

Autoclaving of Biomedical Waste

Autoclave treats the bio-medical waste through the mechanism of disinfection. The biomedical waste was subjected to following temperature and pressure based on its residence time:

- If the autoclave residence time is not less than 60 minutes, the temperature should not be less than 121 °C with the pressure of 15 pounds per square inch (psi).
- If the autoclave residence time is not less than 45 minutes, the temperature should not be less than 135 °C with the pressure of 31 pounds per square inch (psi).
- If the autoclave residence time is not less than 30 minutes, the temperature should not be less than 149 °C with the pressure of 52 pounds per square inch (psi).

While operating a gravity flow autoclave, biomedical waste is subjected to all three condition, whereas in vacuum autoclave, the biomedical waste is first subjected to one pre-vacuum autoclave (minimum) to purge the autoclave of all air. Succeeding this first and second conditions are applied. Bacillus stearothermophilus spore dials or spore strips with at least 1 × 104 spores per ml.is used as biological indicator of autoclave. The operating conditions of autoclave include a residence time less than of 30 minutes, temperature less than 121 °C or a pressure must be less than 15 psi. On reaching certain temperature, the chemical indicator strip/tape changes colour that indicates the attainment of specific temperature. It may be essential to use more than one strip at various locations on the waste package to ensure the effectively autoclaving of inner content of the waste in the package.

Biomedical Liquid Waste

Before disposing the liquid form of biomedical waste into the sewer, it must be treated. Pathological waste after being treated with chemical disinfectants are flushed into the sewage system. Likewise, the chemical waste is neutralized with suitable reagents and then either flushed or treated in the sewage treatment plant. Mostly they are neutralized and dumped in sewer network. Highly skilled operators are required for this technique as it involves handling of hazardous substances. The biomedical waste effluent generated from the various source should conform to the following limits shown in Table below. Environment (Protection) Act, 1986 prescribes the discharge limits of these waste into public sewers.

Table: Disposal standard for biomedical waste.

Parameters Permissible limits	
pH	6.5-9.0
Suspended solids	100 mg/L
Oil and grease	10 mg/L
Biochemical Oxygen Demand (BOD)	30 mg/L
Chemical Oxygen Demand (COD)	250 mg/L

Microwave Treatment

Microwave treatment uses a frequency and wavelength of 2450 MHz and 12.24 cm, respectively for the destruction of microorganisms. The infectious contaminants in water with biomedical waste are destroyed by heat conduction when it is rapidly heated by the microwaves. By bacteriological and biological tests, the efficiency of the microwave disinfection was ensured regularly. The biomedical waste is evenly heated to a temperature of 97-100 °C by means of microwaves in treatment chamber. Treatment of biomedical waste by microwaving can be carried out in the source itself. No shredding is required for microwave treatment of waste.

Most infectious wastes except body parts, human organs, infected animals carcasses and metal objects are suitable for treatment by microwave technique. This method shows good disinfection competence with good waste shrinking capacity. Similar to incineration this method also involves high operating costs. It is an eco-friendly process with potential operation and maintenance problems.

Deep Burial

Deep burial process is done in pits or trench of about 2 meters deep. The pits are half filled with waste, 50 cm soil and then with waste. The pits are covered with galvanized iron / wire meshes. When wastes are added to the pit, a layer of 10 cm of soil shall be added to enclose the waste. The deep burial site should be impermeable with no shallow well in the nearby area. The pits should be away from the habitation to avoid infection to surface or ground water. The site selected should not be a flooding or eroding zone and should be approved by the authority.

Inertization

Assimilation of waste with cement and other substances before disposal is called inertization process. This decreases the risk of entry of toxic substances into the surface or groundwater. A typical percentage of the mixture is 65% pharmaceutical waste, 15 % cement and 5 %water. A homogenous mass is created and cubes or pellets are produced and then stored. This process is economical and not suitable for infectious waste.

Table below shows the treatment and disposal method of the different categories of biomedical waste. The process such as incineration, deep burial, disinfecting process and municipal landfill disposal will be carried out. Category 1, 2, 3, 5 & 6 can be incinerated. Disinfecting process includes chemical treatment, autoclaving, microwaving and mutilation shredding was carried for waste under category 3, 4, 6, 7, 8 and 10. Category 1 and 2 can be disposed off by deep burial. Category 9 waste was disposed by municipal landfill.

Table: Treatment and disposal of biomedical waste.

Waste Category	Treatment And Disposal Option
Category No. 1	Incineration/deep burial.
Category No. 2	Incineration/deep burial.
Category No. 3	Local autoclaving/ microwaving/incineration.
Category No. 4	Disinfecting (chemical treatment/autoclaving/microwaving and mutilation/shredding).

Category No. 5	Incineration/destruction and drugs disposal in secured landfills.
Category No. 6	Incineration/ autoclaving/microwaving.
Category No. 7	Disinfecting by chemical treatment/autoclaving/microwaving and mutilation/shredding).
Category No. 8	Disinfecting by chemical treatment and discarding it into drains.
Category No. 9	Disposal in municipal landfill.
Category No. 10	Chemical treatment and release into drains for liquids and protected landfill for solids.

Municipal Solid Waste

Municipal Solid Waste (MSW) can be defined as solid waste which includes all domestic refuse and non-hazardous wastes such as commercial and institutional wastes, street sweepings and construction debris. In some countries the solid wastes management system also handles human wastes such as night-soil, ashes from incinerators, septic tank sludge and sludge from sewage treatment plants. If these wastes manifest hazardous characteristics they should be treated as hazardous wastes.

Waste management practices differ for developed and developing nations, for urban and rural areas, and for residential and industrial producers. MSW is thus seen as primarily coming from households but also includes wastes from offices, hotels, shopping complexes/shops, schools, institutions, and from municipal services such as street cleaning and maintenance of recreational areas. Residential and commercial types of solid waste include food wastes, paper, cardboard, plastics, textiles, glass, metals, and ashes, special wastes like bulky items, consumer electronics, batteries, oil, tires and household hazardous wastes. Institutions types of solid waste include paper, cardboard, plastics, wood, food wastes, glass, metals, special wastes, hazardous wastes. Municipal services types of solid waste include Street sweepings, landscape and tree trimmings, general wastes from parks, beaches, and other recreational areas. Therefore, the major types of MSW are food wastes, paper, plastic, rags, metal and glass, with some hazardous household wastes such as electric light bulbs, batteries, discarded medicines and automotive parts.

In recent years the volume of waste has been increasing at an alarming rate, posing a formidable challenge to governments. The complexities and enormity of the challenges become evident when considering other waste types to be managed and these include industrial solid waste, municipal wastewater, industrial wastewater, storm water and hazardous waste. Often, different government agencies are mandated to manage different waste sectors. This fragmented approach to waste management, coupled with a lack of clear definition and delineation of the different waste types, makes an assessment of current waste management practices in most countries difficult.

Waste Management Strategies used in Municipal Solid Waste Management

Waste management is the collection, transport, processing, recycling or disposal, and monitoring of waste materials. Operations strategy can be viewed as part of a planning process that coordinates operational goals with those of the larger organization. Since the goals of the larger organization change over time, the operations structure must be designed to anticipate future needs. The operations capabilities of a firm can be viewed as a portfolio best suited to adapt to the changing product and service needs of a firm's customers.

The costs for solid waste management are high especially for collection, transportation, treatment and disposal, which are largely borne by city councils. Methods of collection of waste are either door-to-door or using containers or communal bins. All medium and large cities have administrative structures for providing collection services but often, cities in developing countries use non-compaction trucks for daily collection, with a few cities using compaction trucks and hauling trucks. The most common municipal waste management practices include: recycling/recovery, composting, incineration and land filling/open dumping. The operations strategy is a very important tool in the solid waste management practices and processes.

MSW may contain the following materials, which are considered recyclables: ferrous and non-ferrous metals, construction debris, scrap tires, paper/cardboard, plastics, textiles (including cloth and leather), glass, wood/timber, animal bones/feathers, waste oil and grease, cinders/ashes. In the middle-to-low-income cities, there exists a long-standing practice of informal source separation and recycling of materials. This has led to the development of enterprises for the gathering, trading and reprocessing of materials. For example Mukuru Recycling project which started in 1991 to help men and women scavengers sell recyclable waste to industries. The national ministries support waste recovery and recycling activities at city level although many of these are family businesses. However, since industries would only be interested to use recycled materials when they cost less than the virgin materials, the practice of recycling is so market-driven that recycling has become selective. The disposal of those unselected recyclables remains a problem.

Informal waste separation or waste picking takes place in three ways: At source - this is in large urban areas, e.g., commercial areas or residential areas with apartments/high-rise buildings for high income earners. Here waste pickers sort out the waste before the authorized collection vehicle arrives. During collection, when the collectors segregate recyclable materials during loading and store them inside the truck or on the sides of the vehicles. At the disposal site – where the waste pickers often live on or near the dumps. However, they risk the danger of potential slides and fires. While waste picking means survival for waste pickers the methods of uncontrolled waste picking can reduce the efficiency of the formal collection system and can be detrimental to health due to exposure to biological pathogens.

Composting is not well practiced. Waste materials that are organic in nature, such as plant material, food scraps, and paper products, can be recycled using biological composting and digestion processes to decompose the organic matter. The resulting organic material is then recycled as mulch or compost for agricultural or landscaping purposes Household organic wastes, including wastes from the restaurants, are often collected for animal feed. But these are either not working or are not operating at full capacity for a number of reasons, such as: High operating and maintenance costs, poor maintenance and operation of facilities, Incomplete separation of non-compostable, such as, plastics and glass, high cost of compost compared to commercial fertilizers.

Another waste treatment method that is practiced is incineration where 90 percent of non-recyclable municipal solid waste is incinerated. Final disposal of waste is at landfills where 10 percent of non-recyclable municipal solid waste is deposited. Singapore has four government-owned and operated incinerators for the disposal of solid waste that is not recycled. However, controversy remains over the soundness of incineration as a waste treatment technology because of greenhouse gas emissions from incinerators. Incineration has been completely banned under the new law on solid waste management. The practice of informal incineration or open burning is, however, still prevalent, not only in the rural areas where waste collection is rare but also in peri-urban and urban areas.

The popular meaning of 'recycling' in most developed countries refers to the widespread collection and reuse of everyday waste materials such as empty beverage containers. These are collected and sorted into common types so that the raw materials from which the items are made can be reprocessed into new products. Material for recycling may be collected separately from general waste using dedicated bins and collection vehicles, or sorted directly from mixed waste streams.

Landfills are generally the cheapest and most common disposal method for municipal solid waste. Disposing of waste in a landfill involves burying the waste, and this remains a common practice in most countries. Landfills were often established in abandoned or unused quarries, mining voids or borrow pits. A properly designed and well-managed landfill can be a hygienic and relatively inexpensive method of disposing of waste materials. Older, poorly designed or poorly managed landfills can create a number of adverse environmental impacts such as wind-blown litter, attraction of vermin, and generation of liquid leachate. An exception is a large city like Singapore, which faces rising disposal costs due to exhaustion of traditional disposal sites, stricter environmental controls and greater waste quantities, thus requiring other methods like incineration to reduce the volume of waste for final disposal. In the other developing countries, open dumping is the common practice, i.e., municipal solid waste is dumped on swamplands and low-lying areas, which are eventually reclaimed for development. The problems associated with landfills, even with those that are clay-lined, include high water table, groundwater contamination and gas migration.

Incineration is a disposal method in which solid organic wastes are subjected to combustion so as to convert them into residue and gaseous products. This method is useful for disposal of residue of both solid waste management and solid residue from waste water management. This process reduces the volumes of solid waste to 20 to 30 percent of the original volume. Incineration and other high temperature waste treatment systems are sometimes described as "thermal treatment". Incinerators convert waste materials into heat, gas, steam and ash. Incineration is common in countries such as Japan where land is more scarce, as these facilities generally do not require as much area as landfills. Waste-to-energy (WtE) or energy-from-waste (EfW) are broad terms for facilities that burn waste in a furnace or boiler to generate heat, steam and/or electricity.

Formulation of the Municipal Solid Waste Management Operations Strategy

Operations strategy is the "HOW" in any corporate and market strategy. Operations strategy is no longer a tool for continuous improvement and sustainable competitive advantage in the manufacturing sector only, since it can be now applied in the service industry and public organizations. The operation strategies used in solid waste management can be modeled using a process chart as follows:

From the model, the formulation of organizational strategy must be done by the CEO and the employees through selected committees. The formulation of organizational strategy should be followed by setting of Annual Objectives in Solid Waste Management. The annual objectives includes to: improve public health of the people; improve the environment; and maintain public cleanliness in order to keep public places aesthetically acceptable; by ensuring the proper storage, collection, transportation, safe treatment and disposal of solid waste. This driven by the annual departmental objectives designed according to the department of environment's major mandates. They are derived from annual departmental objectives especially by the departmental heads and the employees. This is operational Level (origination of Annual Objectives in Solid Waste Management). The policies adequately support the institutions strategic plan: the departmental organizational structure support implementation of strategy; and the procedures/regulations followed by the departments are supportive of change implementation. This should be documented in the current strategic plan.

There are so many factors that can enable an organization to take a fresh look at its operations Strategy. The different factors that impact on the operations strategy are: most managers felt that the emergence of aggressive and highly competent competitors, demanding and environmentally conscious customers. Other secondary factors include: advances in production and information technology, global business operations, business process re-engineering techniques and the enormous opportunities for operational efficiencies and economies. The external environment of the municipality's trends in the market; changes in customer wants and expectations (demanding speed of delivery, high quality, and low price); Identifying the company's strengths, special skills of workers, such as expertise in providing customized services or knowledge of information technology; the trends in the political environment changes in the political climate—local, national, and international) and forming partnerships with international firms.

Investment on scanning the environment of operation before developing the solid waste management operations strategy is important. As one invests on time and effort in implementing its operations strategy after the environmental scanning and formulation of operations strategy, there is need to invest in its implementation. There are a number of challenges which needs to be managed during the implementation of operations strategy with respect to solid waste management. The factors considered to have a strongly effect on the municipal solid waste strategy implementation include: Inability to formulate and implement sound solid waste management policies, Inadequate treatment and disposal of solid waste; Inadequate landfill disposal, Poor formulation and enforcement of laws and regulations relating to solid waste management; Poor services for solid waste collection and Limited utilization of recycling activities; the Inability to regulate and monitor the activities of all generators of solid waste, and Poor transportation services provision. The secondary challenges include: The deficient accounting systems ; Dumping along the back lanes and street corners within the city and its suburbs; High risk of environmental pollution; Inadequate service coverage and operational inefficiencies of service; Large amounts of uncollected solid waste; Lack of adequate and appropriate staff ; Inadequate management of hazardous and healthcare waste; No controls from the management side to prevent toxic and even hazardous waste; Very inadequate billing systems.

References

- Introductory-chapter-introduction-to-hazardous-waste-management, management-of-hazardous-wastes: intechopen.com, Retrieved 29, April 2020
- Biomedical-waste-and-its-management-308294194: researchgate.net, Retrieved 20, July 2020
- Types-of-radioactive-waste: conserve-energy-future.com, Retrieved 07, May 2020
- The-generation-composition-collection-treatment-and-disposal-system-and-impact-of-e-waste: intechopen.com, Retrieved 13, June 2020
- Operationalising-municipal-solid-waste-management, integrated-waste-management-volume-ii: intechopen.com, Retrieved 29, April 2020

Chapter 3

Waste Collection

Any garbage that has been discarded by the public falls under the category of solid waste. This chapter encompasses processes and techniques related to solid waste collection such as collection components, waste collection, collection operation, waste collection system design, storage, record keeping, etc. These topics are crucial for a complete understanding of the subject.

Collection Components

Waste collection does not mean merely the gathering of wastes, and the process includes, as well, the transporting of wastes to transfer stations and/or disposal sites. To elaborate, the factors that influence the waste collection system include the following (EPA, 1989 and Ali, et al., 1999):

(i) Collection points: These affect such collection system componentsas crew size and storage, which ultimately control the cost of collection. Note that the collection points depend on locality and may be residential, commercial or industrial.

(ii) Collection frequency: Climatic conditions and requirements of a locality aswell as containers and costs determine the collection frequency. In hot and humid climates, for example, solid wastes must be collected at least twice a week, as the decomposing solid wastes produce bad odour and leachate.

And, as residential wastes usually contain food wastesandother putrescible (rotting) material, frequent collection is desirable for health and aesthetic reasons. Besides climates, the quality of solid waste containers on site also determines the collection frequency. For instance, while sealed or closed containers allow collection frequency up to three days, open and unsealed containers may require daily collection. Collection efficiency largely depends on the demography of the area (such as income groups, community, etc.), where collection takes place. While deciding collection frequency, therefore, you must consider the following:

- cost, e.g., optimal collection frequency reduces the cost as it involves fewer trucks, employees and reduction in total route distance;
- storage space, e.g., less frequent collection may require more storage space in the locality;
- sanitation, e.g., frequentcollectionreducesconcernsabouthealth, safety and nuisance associated with stored refuse.

(iii) Storage containers: Proper container selection can save collection energy, increase the speed of collection and reduce crew size. Most importantly, containers should be functional for the amount and type of materials and collection vehicles used.Containers should also be durable, easy to handle, and economical, as well as resistant to corrosion, weather and animals. In residential areas, where refuse is collected manually, standardised metal or plastic containers are typically required for waste storage. When mechanised collection systems are used, containers are specifically designed to fit the truck-mounted loading mechanisms. While evaluating residential wastecontainers, consider the following:

- efficiency, i.e., thecontainersshouldhelpmaximisetheoverall collection efficiency.
- convenience, i.e., the containers must be easily manageable both for residents and collection crew.
- compatibility, i.e., thecontainersmustbecompatiblewithcollection equipment.
- public health and safety, i.e., the containers should be securely covered and stored.
- ownership, i.e., the municipal ownership must guarantee compatibility with collection equipment.

(iv) Collection crew: The optimum crew size for a community depends on labour and equipment costs, collection methods and route characteristics. The size of the collection crew also depends on the size and type of collection vehicle used, space between the houses, waste generation rate and collection frequency. For example, increase in waste generation rate and quantity of wastes collected per stop due to less frequent collection result in a bigger crew size. Note also that the collection vehicle could be a motorised vehicle, a pushcart or a trailer towed by a suitable prime mover (tractor, etc.). It is possible to adjust the ratio of collectors to collection vehicles such that the crew idle time is minimised. However, it is not easy to implement this measure, as it may result in an overlap in the crew collection and truck idle time. An effective collection crew size and proper workforce management can influence the productivity of the collection system. The crew size, in essence, can have a great effect on overall collection costs. However, with increase in collection costs, the trend in recent years is towards:

- decrease in the frequency of collection.
- increase in the dependence on residents to sort waste materials; increase in the degree of automation used in collection.
- This trend has, in fact, contributed to smaller crews in municipalities.

(v) Collection route: The collection programme must consider the route that is efficient for collection. An efficient routing of collection vehicles helps decrease costs by reducing the labour expended for collection. Proper planning of collection route also helps conserve energy and minimise working hours and vehicle fuel consumption. It is necessary therefore to develop detailed route configurations and collection schedules for the selected collection system. The size of each route, however, depends on the amount of waste collected per stop, distance between stops, loading time and traffic conditions. Barriers, such as railroad, embankments, rivers and roads with heavy traffic, can be considered to divide route territories. Routing (network) analyses and planning can:

- increase the likelihood of all streets being serviced equally and consistently;
- help supervisors locate or track crews quickly;
- provide optimal routes that can be tested against driver judgement and experience.

(vi) Transfer station: A transfer station is an intermediate station between final disposal option and collection point in order to increase the efficiency of the system, as collection vehicles and crew remain closer to routes. If the disposal site is far from the collection area, it isjustifiable to have a transfer station, where smaller collection vehicles transfer their loads to larger vehicles, which then haul the waste long distances. In some instances, the transfer station serves as a pre- processing point, where wastes are dewatered, scooped or compressed. A centralised sorting and recovery of recyclable materials are also carried out at transfer stations (EPA, 1989). The unit cost of hauling solid wastes from a collection area to a transfer station and then to a disposal site decreases, as the size of the collection vehicle increases. This is duetovarious reasons such as the following:

- Labour costs remain constant;
- The ratio of payload to vehicle load increases with vehicle size;
- The waiting time, unloading time, idle time at traffic lights and driver rest period are constant, regardless of the collection vehicle size.
- Efficiency: Do the services help minimise the cost per household?
- Effectiveness: Do the services satisfy the community needs?
- Equity: Do the services address equally the concerns of all social and demographic groups?
- Reliability: Do the services ensure consistency?
- Safety and environmental impact: Do the services ensure safety of workers, public health and protection of the environment?

Note also that various management arrangements, ranging from municipal services to franchised services and under various forms of contracts are, typically, in vogue for waste collection. One of the critical decisions to be made at the planning stage, therefore, is as to who – the public or private agencies – operates the collection system, though the final decision depends on the existing conditions and options for the local decision-makers (EPA, 1989).

Waste Collection

Waste collection is a part of the process of waste management. It is the transfer of solid waste from the point of use and disposal to the point of treatment or landfill. Waste collection also includes the curbside collection of recyclable materials that technically are not waste, as part of a municipal landfill diversion program.

A waste collection vehicle in Sakon Nakhon, Thailand.

Manual waste collection in Bukit Batok West, Singapore.

Waste on a sidewalk for collection, bagged and stickered - in Dublin, Ireland.

Household Waste

Household waste in economically developed countries will generally be left in waste containers or recycling bins prior to collection by a waste collector using a waste collection vehicle.

However, in many developing countries, such as Mexico and Egypt, waste left in bins or bags at the side of the road will not be removed unless residents interact with the waste collectors.

Mexico City residents must haul their trash to a waste collection vehicle which makes frequent stops around each neighborhood. The waste collectors will indicate their readiness by ringing a distinctive bell and possibly shouting. Residents line up and hand their trash container to the waste collector. A tip may be expected in some neighborhoods. Private contracted waste collectors may circulate in the same neighborhoods as many as five times per day, pushing a cart with a waste container, ringing a bell and shouting to announce their presence. These private contractors are not paid a salary, and survive only on the tips they receive. Later, they meet up with a waste collection vehicle to deposit their accumulated waste.

The waste collection vehicle will often take the waste to a transfer station where it will be loaded up into a larger vehicle and sent to either a landfill or alternative waste treatment facility.

Commercial Waste

Waste collection considerations include type and size of bins, positioning of the bins, and how often bins are to be serviced. Overfilled bins result in rubbish falling out while being tipped. Hazardous rubbish like empty petrol cans can cause fires igniting other trash when the truck compactor is operating. Bins may be locked or stored in secure areas to avoid having non-paying parties placing rubbish in the bin.

Kerbside Collection

Kerbside collection in Canberra, Australia

Kerbside collection, orcurbside collection, is a service provided to households, typically in urban and suburban areas of removing household waste. It is usually accomplished by personnel using purpose built vehicles to pick up household waste in containers acceptable to or prescribed by the municipality.

History

Prior to the 20th century the amount of waste generated by a household was relatively small. Household wastes were often simply thrown out the window, buried in the garden or deposited in outhouses. When human concentrations became more dense, waste collectors, called nightmen or gong farmers were hired to collect the night soil from pail closets, performing their duties only at night (hence the name). Meanwhile, disposing of refuse became a problem wherever cities grew. Often refuse was placed in unusable areas just outside the city, such as wetlands and tidal zones. One example is London, which from Roman times disposed of its refuse outside the London Wall beside the River Thames. Another example is 1830s Manhattan, where thousands of hogs were permitted to roam the streets and eat garbage. A small industry developed as "swill children" collected kitchen refuse to sell for pig feed and the rag and bone man traded goods for bones (used for glue) and rags (essential for paper manufacture prior to the invention of wood pulping). Later, in the late nineteenth century, trash was fed to swine in industrial.

As sanitation engineering came to be practised beginning in the mid-19th century and human waste was conveyed from the home in pipes, the gong farmer was replaced by the municipal trash collector as there remained growing amounts of household refuse, including fly ash from coal, which was burnt for home heating. In Paris, the rag and bone man worked side by side with the municipal bin man, though reluctantly: in 1884, Eugène Poubelle introduced the first integrated kerbside collection and recycling system, requiring residents to separate their waste into perishable items, paper and cloth, and crockery and shells. He also established rules for how private collectors and city workers should cooperate and he developed standard dimensions for refuse containers: his name in France is now synonymous with the garbage can. Under Poubelle, food waste and other organics collected in Paris were transported to nearby Saint Ouen where they were composted. This continued well into the 20th century when plastics began to contaminate the waste stream.

From the late-19th century to the mid-20th century, more or less consistent with the rise of consumables and disposable products municipalities began to pass anti-dumping ordinances and introduce kerbside collection. Residents were required to use a variety of refuse containers to facilitate kerbside collection but the main type was a variation of Poubelle's metal garbage container. It was not until the late 1960s that the green bin bag was introduced by Glad. Later, as waste management practices were introduced with the aim of reducing landfill impacts, a range of container types, mostly made of durable plastic, came to be introduced to facilitate the proper diversion of the waste stream. Such containers include blue boxes, green bins and wheelie bins or MGBs.

Over time, waste collection vehicles gradually increased in size from the hand pushed tip cart or English dust cart, a name by which these vehicles are still referred, to large compactor trucks.

Waste Management and Resource Recovery

Glass for collection in Edinburgh, Scotland.

Kerbside collection is today often referred to as a strategy of local authorities to collect recyclable items from the consumer. Kerbside collection is considered a low-risk strategy to reduce waste volumes and increase recycling rates. Materials are typically collected in large bins, coloured bags, or small open plastic tubs, specifically designated for content.

Discarded Christmas trees awaiting pickup in the San Fernando Valley.

Recyclable materials that may be separately collected from municipal waste include:

Biodegradable waste component

- Green waste
- Kitchen waste

Recyclable materials, depending on location

- Office paper
- Newsprint
- Paperboard
- Corrugated fiberboard
- Plastics (#1 PET, #2 HDPE natural and colored, #3 PVC narrow-necked containers, #4 LDPE, #5 PP, #6 Polystyrene (however not EXPANDED polystyrene, an example of recyclable polystyrene may be a yoghurt pot) #7 other mixed resin plastics)
- Glass
- Copper
- Aluminum
- Steel and Tinplate
- Co-mingled recyclables- can be sorted by a clean materials recovery facility

In Somerville, MA all accepted paper, glass, plastic, and metal recycling is picked up from a single bin.

Kerbside collection of recyclable resources is aimed to recover purer waste streams with higher market value than by other collection methods. If the household incorrectly separates the recyclable elements, the load may have to be put to landfill if it is deemed to be contaminated.

Kerbside collection and household recycling schemes are also being used as tools by local authorities to increase the public's awareness of their waste production.

Kerbside collection is commonly considered to be completely environmentally friendly. This may not necessarily be the case as it leads to an increased number of waste collection vehicles on the road, in themselves contributing to global warming through exhaust emissions until the time of their conversion to clean energy.

New and emerging waste treatment technologies such as mechanical biological treatment may offer an alternative to kerbside collection through automated separation of waste in recycling factories.

Usage by Country

Canada

Calgary, Alberta has adopted "Curbside" Recycling and uses blue bins. The blue cart programme accepts all types of recycables, including plastics 1-7. It is picked up weekly for the cost of $8.00 per month. This programme is mandatory.

In 1981 Resource Integration Systems (RIS) in collaboration with Laidlaw International tested the first blue box recycling system on 1500 homes in Kitchener, Ontario. Due to the success of the project the City of Kitchener put out a contract for public bid in 1984 for a recycling system city wide. Laidlaw won the bid and continued with the popular blue box recycling system. Today hundreds of cities around the world use the blue box system or a similar variation.

Halifax Regional Municipality (HRM) in Nova Scotia, Canada, with a population of about 375, 000, has one of the most complex kerbside collection programmes in North America. Based on the green cart, it requires residents to self-sort refuse and place different types at the kerb on alternating weeks. As shown in the photo at left, week 1 would see the green cart and optional orange bags used for kitchen waste and other organics such as yard waste. Week 2 would permit non-recoverable waste in garbage bags or cans. Blue bags are used for paper, plastic and metal containers. Together with used grocery bags containing newspapers, they may be placed on the kerb either week. In summer, the green cart is emptied weekly due to the prevalence of flies. HRM has achieved a diversion rate of approximately 60 percent by this method.

Many Canadian municipalities use "green bins" for kerbside recycling. Others, such as Moncton, use wet/dry waste separation and recovery programmes.

New Zealand

Kerbside collection bins in Dunedin, New Zealand. The yellow-liddied wheelie bin is for non-glass recyclables, and the blue bin is for glass. The two bins are collected on alternating weeks. Official council bags are used for general household waste, and are collected weekly.

In New Zealand, kerbside collection of general refuse and recycling, and in some areas organic waste, is the responsibility of the local city or district council, or private contractors. Practices and collection methods vary widely from council to council and company to company. Some examples of collection are:

- Auckland City Council: Two 240-litre wheelie bins are supplied: a red-lidded bin for general refuse, collected weekly, and a blue-lidded bin for recyclables, collected fortnightly.
- Christchurch City Council: Three wheelie bins are supplied: a 140-litre red-lidded bin for general refuse, a 240-litre yellow-lidded bin for recyclables, and an 80-litre green-lidded

bin for organic waste. The organic waste bins are collected weekly, while the recyclables and general refuse bins are collected on alternating weeks.

- Hamilton City Council and Hutt City Council: A 45-litre bin is supplies for recyclables, collected weekly. General refuse is collected weekly using user-pays official council bags.
- Dunedin City Council, Palmerston North City Council and Wellington City Council: Two bins are supplied. A 45-litre or 70-litre bin for glass, and an 80-litre or 240-litre wheelie bin for non-glass recyclables. These two bins are collected on alternating weeks. General refuse is collected weekly using user-pays official council bags.
- Rodney District Council: A 45-litre bin is supplies for recyclables, collected weekly. There is no council collection of general waste, and all general waste collection is carried out by independent companies.
- Taupo District Council: A 45-litre bin is supplies for recyclables, collected weekly. General refuse is collected weekly using user-pays system of orange tags - one orange tag is to be placed on a standard rubbish bag up to 60 litres capacity, or half an orange sticker can be placed on two supermarket bags tied together.
- Upper Hutt City Council: Recycling is to be placed in plastic bags, with paper and cardboard collected in the first week, and plastic, metal and glass in the second week. General refuse is collected weekly using user-pays official council bags.
- Waitakere City Council: A 140-litre wheelie bin is provided for recyclables, collected fortnightly. General refuse is collected weekly using user-pays official council bags.

By 1996 the New Zealand cities of Auckland, Waitakere, North Shore and Lower Hutt had kerbside recycling bins available. In New Plymouth, Wanganui and Upper Hutt recyclable material was collected if placed in suitable bags. By 2007 73% of New Zealanders had access to kerbside recycling.

Kerbside collection of organic waste is carried out by the Mackenzie District Council and the Timaru District Council. Christchurch City Council is introducing the system to their kerbside collection. Other councils are carrying out trials.

United Kingdom

In the United Kingdom, the Household Waste Recycling Act 2003 requires local authorities to provide every household with a separate collection of at least two types of recyclable materials by 2010. There has been criticism in the difference of schemes used in the country such as the colour of bins, whether they are bins boxes or bags, and also the fact that clutter roads and how the additional trucks and collections needed have carbon dioxide emissions too. Some find the colour differences confusing, and people want a national scheme. A typical example is to compare two neighbouring councils in greater Manchester, Bury council and Salford. Bury uses blue for cans, plastic and glass, green for paper and cardboard and brown for garden waste. Salford uses blue for paper and card, brown for cans plastic and glass and pink for garden waste. Most councils use grey or black for general waste, with a few exceptions such as Liverpool, which uses purple for general waste, a colour used by no other council.

Another controversial issue in the uk is the frequency of the waste collections. To save money, many councils are cutting the frequency of both general waste and recyclables collections. This has led to problems from larger families, and has led to overflowing and fly tipping. For example, previously, Bury Council collected general waste once a week and recyclables fortnightly. This has now changed to fortnightly for general waste and monthly (every 4 weeks) collection of recyclables.

A few councils are using "forced" recycling, by replacing the large, 240l general waste bin with a smaller 180l or 140l bin, and using the old 240l one for recyclables. This may be made worse by fortnightly collections of the "small" bin, and strict rules such as "No extra bags will be taken" and "Bin lids must be fully closed". Stockport Council is a notable user of this scheme. Their recycling rates have risen substantially as a result, but there are usually complaints from families. Trafford council also use a similar scheme, but the small grey bin is emptied every week. In addition, the two named councils, and more, collect food waste together with garden waste, by sending out kitchen caddies and compostable liners. These prevent food waste (including meat) from going to landfill, and to increase the councils recycling rate. The food and garden waste is usually collected weekly or fortnightly, and is taken to an In Vessel composter or Anaerobic digester, where the waste is turned into soil improver for use on local farms.

In the north west, all the glass collected is used within the UK, around half of the plastics and cans are used in the UK; the rest is sent further afield to Europe or China to be made into new products, and paper and paperboard collected is sent to local paper mills to be made into newspapers, tissues, paperboard and office paper. Again some of the paper will be sent further afield.

Some councils only have 3 bins- general, organic and recyclables. This means that plastics, cans and glass go in the same container as paper and cardboard. Although this is much easier for the residents, there is more sorting required, and the paper quality is sometimes of a low grade due to food contamination or shards of glass in the paper, and so this scheme is criticized.

Also, most councils require residents to remove caps from bottles and rinse them out to avoid smells. This is because the lids are made from a different type of plastic (PP) to the bottle (PET/HDPE) - although by collapsing the bottles and folding them over like toothpaste tubes and re-screwing the caps in place enables the volume of bottles to be drastically reduced, thereby increasing the amount of bottles that can be carried in the recycling bins. In fact many bottlers, especially bottled water companies, have now designed their bottles to be collapsible; though this message has not been effectively disseminated to the consumer. A collapsible bottle takes between 25% and 33% of the space a non-collapsed bottle.

Labels are rarely required to be removed, however. This also means that only plastic bottles are recycled. Councils are still trying to make clear that plastic tubs (yogurts, desserts and spreads), bags and cling film cannot be recycled through the kerbside economically. If too much contamination is collected then this results in the whole vehicle load going to landfill at a high cost. Contamination is usually a problem if recyclables are collected in wheelie bins, as the worker can only look at the top; there may be contamination 'hidden' at the bottom. Councils that use many bags and boxes (Edinburgh) suffer from less contamination but are complicated and the loose paper and cardboard, and recycling bags are blown around, and paper can be wet.

Basque Country

In the province of Gipuzkoa, this system is implanted in many towns as Usurbil, Hernani, Oiartzun, Antzuola, Legorreta, Itsasondo, Zaldibia, Anoeta, Alegia, Irura, Zizurkil, Astigarraga, Ordizia, Oñati and Lezo, where the common used name in basque is atez-atekoa, which means *door-by-door*. Due to the big success in this towns, with more than the 80% of the waste recycled, 34 towns in Gipuzkoa are studying to set this system up in 2013, like Arrasate, Bergara, Aretxabaleta, Eskoriatza, Legazpi, Tolosa or Pasaia.

The atez-ate system consists in hanging each kind of rubbish in a hanger outside the house a certain day or days in a week. For example, in Hernani, they have three days to hang their organic rubbish, two days for plastics and metallics, one for paper and one for rejects residuals.

This system started in the town of Usurbil in the year 2009, due to the incinerator of the region of Gipuzkoa was going to build in this town, exactly in the neighborhood of Zubieta. Three years after, the construction of the incinerator was paralyzed by the government of the region, suggesting that the incinerator was a source of contamination and the high cost of the building.

Criticism

This type of collection service is subject to growing criticism.

- The large (Wheelie bin) container encourages the "out of sight" rubbish mentality and invites more rubbish to be disposed.
- The bins and collection trucks are not suited to narrow roads or houses with steep driveways or steps.
- They lock local authorities into capital intensive equipment programmes and multi-national providers.
- Co-mingled recyclables are sometimes not being successfully managed by automated sorting stations and the rates of diversion are low. In some cases, this results in mountains of unsorted recyclables.
- In the UK especially, some councils are sending out at least 4 large bins - residents of smaller houses with no gardens have little space to put them.
- Many use small plastic boxes, bags and lockable outdoor food waste 'caddies' which get blown around and lost, bad for recycling participation.

Collection Operation

Now lets discuss the movement of collection crew in terms of workforce efficiency and collection routes.

Movement of Collection Crew

In cultures such as India, Bangladesh, etc., solid waste collection is assigned to the lowest social group. More often, the collection crew member accepts the job as a temporary position or stopgap arrangement, while looking for other jobs that are considered more respectable.

Apart from this cultural problem, the attitude of some SWM authorities affects collection operation. For example, some authorities still think that the collection of solid waste is mechanical, and therefore, the collection crew does not need any training to acquire special skills. As a result, when a new waste collector starts working, he or she is sent to the field without firm instruction concerning his or her duties, responsibilities and required skills. For an effective collection operation, the collection team must properly be trained. The collection crew and the driver of the collection vehicle must, for example, work as a team, and this is important to maintain the team morale and a sense of social responsibility among these workers.

You must also note that the movement of collection crew, container location and vehicle stopping point affect collection system costs. Figure highlights the distance the collection crew will have to walk, if it were to serve the farthest point first or serve the point closest to the vehicle:

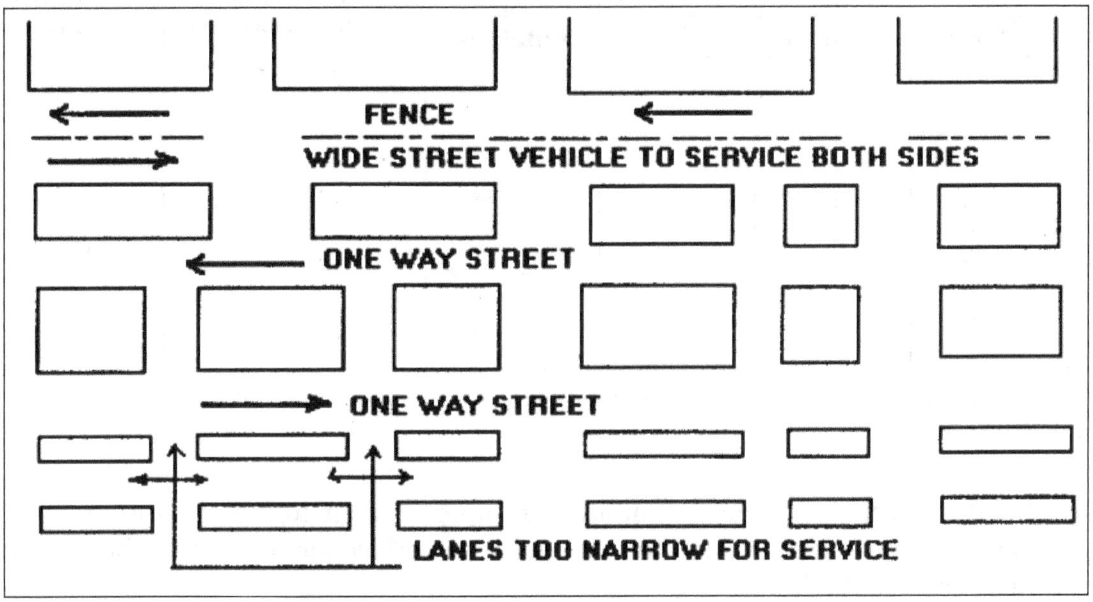

Effect of Container Location and Vehicle Stopping.

The difference may be one or two minutes per collection stop, but it matters with the number of stops the crew will take in a working shift. Multiplying the minutes by the total number of crew working and labour cost depicts the amount of labour hours lost in terms of monetary value.

Generally, familiarity of the crew with the collection area improves efficiency. For example, the driver becomes familiar with the traffic jams, potholes and other obstructions that he or she must avoid. The crew is aware of the location of the containers and the vehicle stops. It is, therefore, important to assign each crew specific areas of responsibility. Working together also establishes an understanding of the strong and weak points of the team members and efficient work sequences. The collection operation must also observe a strict time schedule. Testing of new routes, new gadgets and vehicles is best carried out first in the laboratory and later in a pilot area. Testing of a new sequence using the whole service area could result in disorder and breakdown of the solid waste collection system. Studies show that it takes two hours to recover for every hour of a failed system.

Motion Time Measurement (MTM) Technique

Motion time measurement (MTM) studies are now an integral part of the standard procedure in the development of solid waste collection systems. MTM is a technique to observe and estimate the movement of the collection crew with the help of stopwatches.The results thus gathered are tabulated as shown in Table to determine the best sequence of activities that workers must follow in order to complete a repetitive task in the shortest possible time:

Table:

MTM Study: Determination of Time, Distance and Number on Containers in Collection Route

	Time		**Odometer (Km)**	**Number of Containers**	**Collection time (Minute Second)**	**Trip time to next Station**
	Arrival	**Departure**				
Garage	::	::				
1 Station	::	::				
2 Station	::	::				
.	::	::				
.	::	::				
.	::	::				
20 Station	::	::				
Last Station	::	::				
Disposal	::	::				
Site Total						
Weight	With Load tonne		With Load tonne		With Load tonne	

MTM also helps in deciding the best combination of equipment to maintain a desired level of output, reduce health problems related to the repetitive work sequence and predict the effects of changes in materials handled.

Sophisticated MTM studies involve hidden or open video cameras at different collection stops to record, replay and study the operation sequence of the collection crew. If the crew is conscious of being observed, they tend to work faster and reduce time wastage in unauthorised salvaging and other non- scheduled activities. Once the crew is familiar with the person(s) observing them, it begins to perform more credibly. In studies involving video cameras, therefore, the first two or three hours of observation are often neglected.

Collection Vehicle Routing

Efficient routing and re-routing of solid waste collection vehicles can help decrease costs by reducing the labour expended for collection. Routing procedures usually consist of the following two separate components:

(i) Macro-routing: Macro-routing, also referred to as route-balancing, consists of dividing the total collection area into routes, sized in such a way as to represent a day's collection for each crew. The size of each route depends on the amount of waste collected per stop, distance between stops, loading time and traffic conditions. Barriers, such as railroad embankments, rivers and roads with heavy competing traffic, can be used to divide route territories. As much as possible, the size and shape of route areas should be balanced within the limits imposed by such barriers.

(ii) Micro-routing: Using the results of the macro-routing analysis, micro- routing can define the specific path that each crew and collection vehicle will take each collection day. Results of micro-routing analyses can then be used to readjust macro-routing decisions. Micro-routinganalysesshould also include input and review from experienced collection drivers.

Districting is the other method for collection route design. For larger areas it is not possible for one institution to handle it then the best way is to sub divide the area and MSW collection districting plan can be made.This routing will be successful only when road network integrity is good and the regional proximity has been generated.

The heuristic (i.e., trial and error) route development processisarelatively simple manual approach that applies specific routing patterns to block configurations. The map should show collection, service garage locations, disposal or transfer sites, one-way streets, natural barriers and areas of heavy traffic flow. Routes should then betracedontothetracingpaperusingthe following rules:

- Routes should not be fragmented or overlapping. Each route should be compact, consisting of street segments clustered in the same geographical area.
- Total collection plus hauling time should be reasonably constant for each route in the community.
- The collection route should be started as close to the garage or motor pool as possible, taking into account heavily travelled and one-way streets.
- Heavily travelled streets should not be visited during rush hours.
- In the case of one-way streets, it is best to start the route near the upper end of the street, working down it through the looping process.
- Services on dead-end streets can be considered as services on the street segment that they-intersect, since theycan onlybe collected by passing down that street segment. To keep right turns at a minimum, (in countries where driving is left-oriented) collection from the dead-end streets is done when they are to the left of the truck. They must be collected by walking down, reversing the vehicle or taking a U-turn.
- Waste on a steep hill should be collected, when practical, on both sides of the street while vehicle is moving downhill. This facilitates safe, easy and fast collection. It also lessens wear of vehicle and conserves gas and oil.
- Higher elevations should be at the start of the route.

- For collection from one side of the street at a time, it is generally best to route with many anti-clockwise turns around blocks.
- For collection from both sides of the street at the same time, it is generally best to route with long, straight paths acrossthe grid before looping anti- clockwise.
- For certain block configurations within the route, specific routingpatterns should be applied. (Adapted from American Public Works Association, 1975.)

Based on the above rules, Figure below illustrates a typical collection vehicle routing:

Figure: Collection Vehicle Route

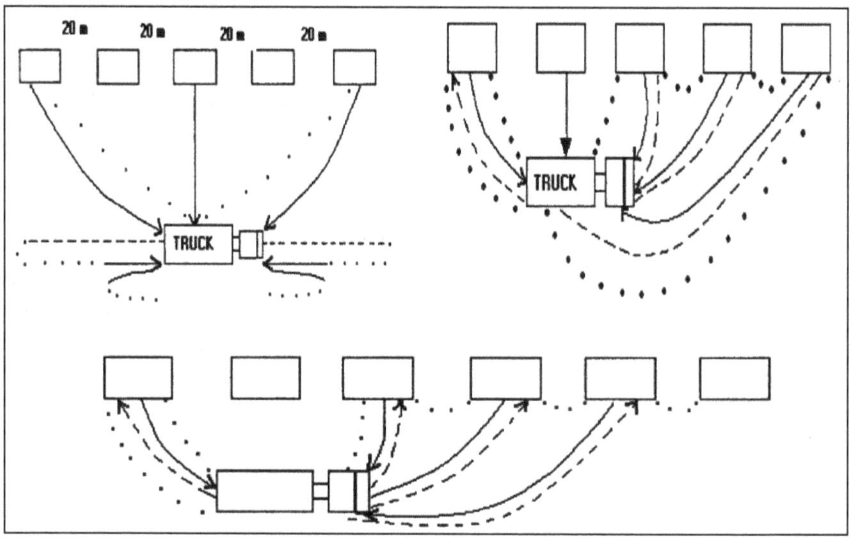

Waste Collection System Design

After we identify appropriate options for collection, equipment and transfer, we must examine the various combinations of these elements to define system -wide alternatives for further analysis. Each should be evaluated for its ability to achieve the identified goals of the collection programme. Economic analysis will usually be a central focus of the system evaluation. This initial evaluation will lead to several iterations, with the differences between the alternatives under consideration becoming more narrowly focused with each round of evaluations (EPA, 1995). After comparing the alternative strategies, the various elements like crew and truck requirement, time requirement and cost involved are calculated. The various formulae used to calculate are:

(i) Number of services/vehicle load (N):

$N = (C \times D) / W$

where, C = Vehicle capacity (m^3) ; D = Waste density (kg/m^3) and W = Waste generation/residence (kg/service).

(ii) Time required collecting one load (E):

$$E = N x L$$

where, L = Loading time/residence, including on-route travel.

(iii) Number of loads/crew/day (n):

The number of loads (n) that eachcrew cancollect in a day canbe estimated based on the workday length (t), and the time spent on administration and breaks (t_1), time for hauling and other travel (t_2) and collection route time (t_3).

Administrative and break time (t_1):

$$t_1 = A + B$$

where, A=Administrativetime(i.e., formeetings, paperwork, unspecified slack time) and B = Time forbreaks and lunch.

Hauling and other travel time (t_2):

$$t_2 = (n x H) - f + G + J$$

where, n = Number of loads/crew/day; H = Time to travel to disposal site, empty truck, and return to route; f = Time to return from site to route; G = Time to travel from staging garage to route and J = Time to return from disposal site to garage.

Time spent on collection route (t_3):

$$t_3 = n x E$$

where variables have been previously defined.

Length of workday (t):

$$t = t_1 + t_2 + t_3$$

where t is defined by work rules and equations A through D are solved to find n.

(iv) Calculation of number of vehicles and crews (K):

$$K = (S x F) / (N x n x M)$$

where, S = Total number of services in the collection area; F = Frequency of collection (numbers/ week) and M =Number of workdays/week.

(v) Calculation of annual vehicle and labour costs: Vehicle costs =Depreciation + Maintenance + Consumables + Overhead + License + Fees + Insurance.

Labour costs =Drivers salary + Crew salaries + Fringe benefits + Indirect labour + Supplies

+ Overhead

Storage: Containers/Collection Vehicles

Waste storage encompasses proper containers to store wastes and efficient transport of wastes without any spillage to transfer stations/disposal sites.

Containers/Storage Bins

The design of an efficient waste collection system requires careful consideration of the type, size and location of containers at the point of generation for storage of wastes until they are collected. While single-family households generally use small containers, residential units, commercial units, institutions and industries require large containers. Smaller containers are usually handled-manually whereas the larger, heavier ones require mechanical handling. The containers may fall under either of the following two categories:

(i) Stationary containers: These are used for contents to be transferred to collection vehicles at the site of storage.

(ii) Hauled containers: These are used for contents to be directly transferred to a processing plant, transfer station or disposal site for emptying before being returned to the storage site.

The desirable characteristics of a well-designed container are low cost, size, weight, shape, resistance to corrosion, water tightness, strength and durability (Phelps, et al., 1995). For example, a container for manual handling by one person should not weigh more than 20 kg, lest it may lead to occupational health hazards such as muscular strain, etc. Containers that weigh more than 20 kg, when full, require two or more crew members to manually load and unload the wastes, and which result in low collection efficiency.

Containers should not have rough or sharp edges, and preferably have a handle and a wheel to facilitate mobility. They should be covered to prevent rainwater from entering (which increases the weight and rate of decomposition of organic materials) into the solid wastes. The container body must be strong enough to resist and discourage stray animals and scavengers from ripping it as well as with stand rough handling by the collection crew and mechanical loading equipment. Containers should be provided with a lifting bar, compatible with the hoisting mechanism of the vehicle. The material used should be light, recyclable, easily moulded and the surface must be smooth and resistant to corrosion. On the one hand, steel and ferrous containers are heavy and subject to corrosion; the rust peels off exposing sharp edges, which could be hazardous to the collection crew. On the other, wooden containers (e.g., bamboo, rattanand wooden baskets) readily absorb and retain moisture and their surfaces are generally rough, irregular and difficult to clean.

Communal Containers

Generally, the containers used for waste storage arecommunal/public containers. Figure below shows a typical communal container, which a compactor collection vehicle can lift and empty mechanically:

The use of communal containers is largely dependent on local culture, tradition and attitudes towards waste. Communal containers may be fixed on the ground (stationary) or movable (hauled).

Movable containers are provided with hoists and tails compatible with lifting mechanism of collection vehicles and such containers have capacities of 1 – 4 m3. The waste management authority must monitor, maintain and upgrade the communal containers. Note that in residential and commercial areas in India, the communal containers are often made of concrete.

Typical Communal Container.

In areas with very high waste generation rates, i.e., rates exceeding two truckloads daily, such as wet markets, large commercial centres and large business establishments, roll-on-roll or hoisted communal containers with capacities of 12 – 20 m3 and a strong superstructure with wheels are used. Normally, the collection vehicle keeps an empty container asareplacement before it hauls the filled container. When a truck is used as a collection vehicle, the use of communal containers may be appropriate.

It is advisable to place the containers 100 – 200 m apart for economic reasons. The communal containers are usually staggered such that the effective distance of 100 m is maintained as shown in Figure:

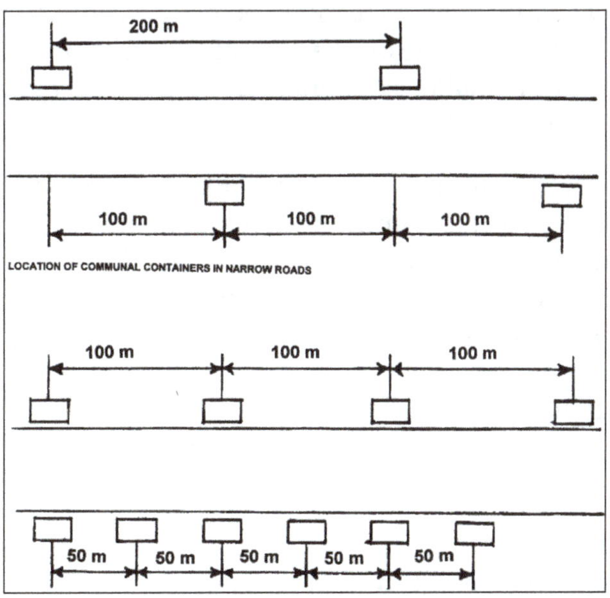

Location of Communal Container.

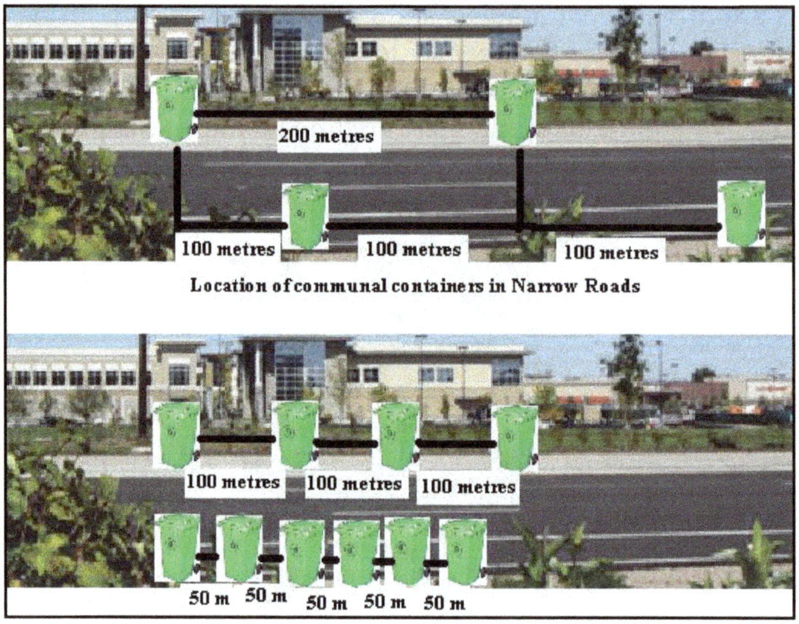

This means that the farthest distance the householder will have to walk is 50 meters. However, in narrow streets with low traffic, where the house owner can readily cross the street, a longer distance is advisable. If the collection vehicle has to stop frequently, say, at every 50 m or so, fuel consumption increases, and this must be avoided.

Disadvantages

The major disadvantage of communal containers is the potential lack of maintenance and upgrading. The residuals and scattered solid wastes emit foul odours, which discourage residents from using the containers properly. In addition, if fixed containers are built below the vehicle level, the collection crew may be held responsible for sweeping and loading the solid wastes into transfer containers before being loaded into the collection vehicle. Sweepingand cleaning the communal containers of residuals obviously impinge on the time of the crew members and take a longer time than if the wastes are placed in smaller containers. As fixed communal containers have higher rates of failure, their use is not advisable.

To overcome the problem of maintaining communal containers, individual residents should maintain their own containers and locate them in designated areas. The communal area must have water and drains to facilitate the cleaning of the containers. This practice has the advantage of reducing the number of collection stops and at the same time maintaining the householder's responsibility for cleaning them. The residents must also be properly educated on the importance of good housekeeping as the containers in the communal area are subject to vandalism. In the main, if communal containers are to be successful, the design of the containers, loading and unloading areas, and collection vehicle accessories should be co-ordinated.

Collection Vehicles

Almost all collections are based on collector and collection crew, which move through the collection service area with a vehicle for collecting the waste material. The collection vehicle selected must be appropriate to the terrain, type and density of waste generation points, the way it travels and type and kind of material (UNEP, 1996). It also depends upon strength, stature and capability of the crew that will work with it. The collection vehicle may be small and simple (e.g., two-wheeled cart pulled by an individual) or large, complex and energy intensive (e.g., rear loading compactor truck). The most commonly used collection vehicle is the dump truck fitted with a hydraulic lifting mechanism. A description of some vehicle types follows:

(i) Small-scale collection and muscle-powered vehicles: These are common vehicles used for waste collection in manycountries and are generally used in rural hilly areas. As Figure illustrates, these can be small rickshaws, carts or wagons pulled by people or animals, and are less expensive, easier to build and maintain compared to other vehicles:

They are suitable for denselypopulated areas with narrow lanes, and squatter settlements, where there is relatively low volume of waste generated. Some drawbacks of these collection vehicles includelimited travel range of the vehicles and weather exposure that affect humans and animals.

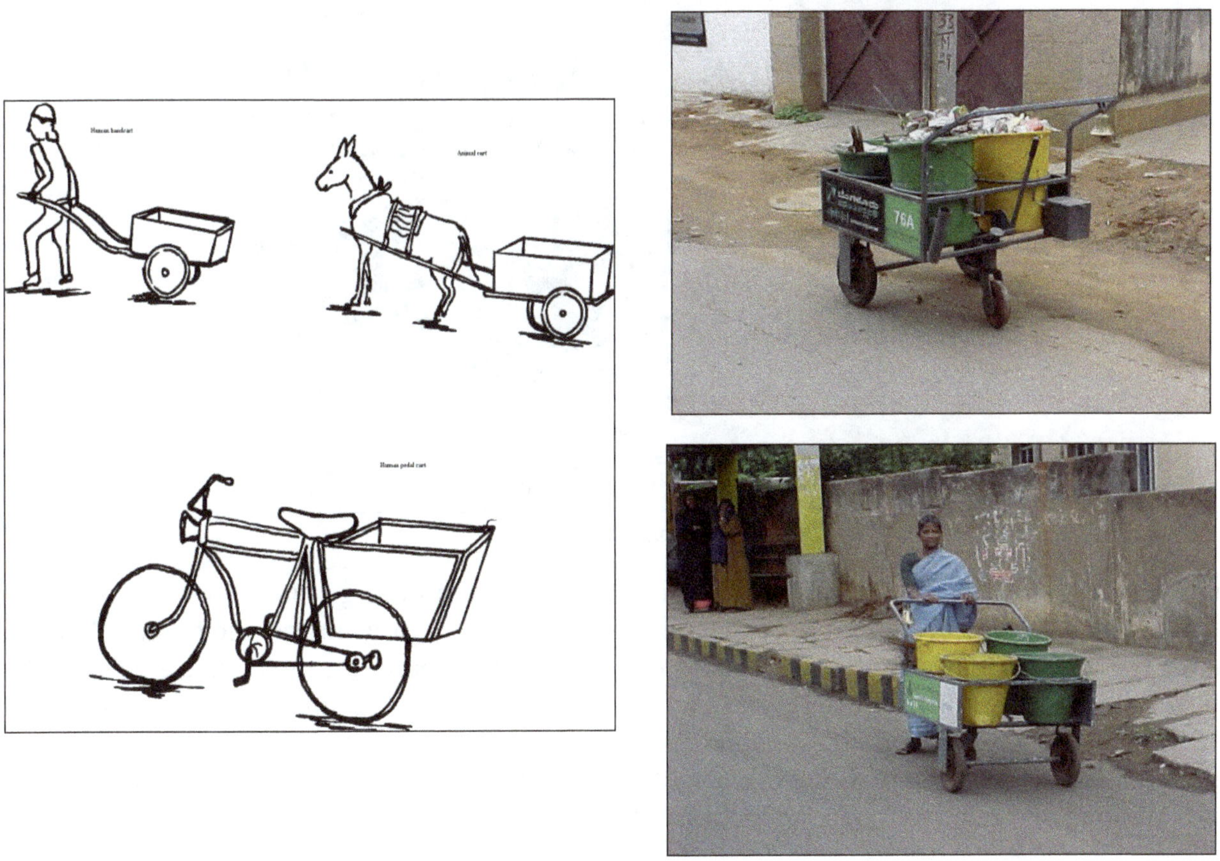

Small-scale Collection Vehicles: An Illustration.

(ii) Non-compactor trucks: Non-compactor trucks are efficient and cost effective in small cities and in areas where wastes tend to be very dense and have little potential for compaction. Figure illustrates a non- compactor truck:

Non-compactor Trucks.

When these trucks are used for waste collection, they need a dumping system to easily discharge the waste. It is generally required to cover the trucks in order to prevent residue flying off or rain soaking the wastes. Trucks with capacities of 10 – 12 m3 are effective, if the distance between the disposal site and the collection area is less than 15 km. If the distance is longer, a potential transfer station closer than 10 km from the collection area is required. Non-compactor trucks are generally used, when labour cost is high. Controlling and operating cost is a deciding factor, when collection routes are long and relatively sparsely populated.

(iii) Compactor truck: Compaction vehicles are more common these days, generally having capacities of 12 – 15 m3 due to limitations imposed by narrow roads. Although the capacity of a compaction vehicle, illustrated in Figure, is similar to that of a dump truck, the weight of solid wastes collected per trip is 2 to 2.5 times larger since the wastes are hydraulically compacted:

Compactor Truck.

The success of waste management depends on the level of segregationat source. One of the examples for best collection method is illustrated in the figure below:

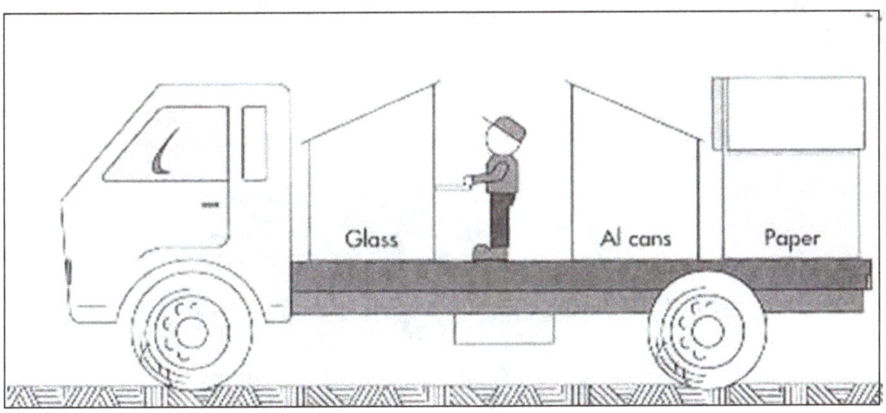

A compactor truck allows waste containers to be emptied into the vehicle from the rear, front or sides and inhibits vectors (of disease) from reaching the waste during collection and transport. It works poorly when wastestream is very dense, wet, collected materials are gritty or abrasive, or when the roads are dusty. The advantages of the compactor collection vehicle include the following:

- containers are uniform, large, covered and relatively visually inoffensive;
- waste is set out in containers so that the crew can pick them up quickly;

- health risk to the collectors and odour on the streets are minimised;
- waste is relatively inaccessible to the waste pickers.

Garbage Truck

Garbage truck or dustcart refers to a truck specially designed to collect municipal solid waste and haul the collected waste to a solid waste treatment facility such as a landfill. Other common names for this type of truck include trash truck in the United States, and rubbish truck, bin wagon, dustbin lorry, bin lorry or bin van elsewhere. Technical names include waste collection vehicle and refuse collection vehicle. These trucks are a common sight in most urban areas.

A Scania front loader.

Major U.S. manufacturers of garbage trucks include Mack and Autocar Trucks. Major manufacturers of garbage truck bodies (not the truck itself) include McNeilus, and Heil.

History

Thornycroft Steam Dust-Cart of 1897 with tipper body.

Wagons and other means had been used for centuries to haul away solid waste. Among the first self-propelled garbage trucks were those ordered by Chiswick District Council from the Thornycroft Steam Wagon and Carriage Company in 1897 described as a steam motor tip-car, a new design of body specific for "the collection of dust and house refuse".

The 1920s saw the first open-topped trucks being used, but due to foul odors and waste falling from the back, covered vehicles soon became more common. These covered trucks were first introduced in more densely populated Europe and then in North America, but were soon used worldwide.

The main difficulty was that the waste collectors needed to lift the waste to shoulder height. The first technique developed in the late 20s to solve this problem was to build round compartments with massive corkscrews that would lift the load and bring it away from the rear. A more efficient model was the development of the hopper in 1929. It used a cable system that could pull waste into the truck.

In 1937, George Dempster invented the *Dempster-Dumpster* system in which wheeled waste containers were mechanically tipped into the truck. His containers were known as Dumpsters, which led to the word dumpster entering the language.

In 1938, the Garwood Load Packer revolutionized the industry when the notion of including a compactor in the truck was implemented. The first primitive compactor could double a truck's capacity. This was made possible by use of a hydraulic press which compacted the contents of the truck periodically.

RS-3 Lightning Rear Steer truck.

1955 saw the Dempster Dumpmaster the first front loader introduced, however they didn't become common until the 1970s. The 1970s also saw the introduction of smaller dumpsters, often known as wheelie bins which were also emptied mechanically. Since that time there has been little dramatic change, although there have been various improvements to the compaction mechanisms in order to improve payload. In the mid-1970s Petersen Industries introduced the first grapple truck for municipal waste collection.

In 1997, Lee Rathbun introduced the *Lightning Rear Steer System*. This system includes an elevated, rear-facing cab for both driving the truck and operating the loader. This configuration allows the operator to follow behind haul trucks and load continuously.

Types of Waste Collection Vehicle

A standard Waste Management Inc. front-loading garbage truck in San Jose, California.

Front Loaders

Front loaders generally service commercial and industrial businesses using large waste containers with lids known as Dumpsters in the US. The truck is equipped with powered forks on the front which the driver carefully aligns with sleeves on the waste container using a joystick or a set of levers. The waste container is then lifted over the truck. Once it gets to the top the container is then flipped upside down and the waste or recyclable material is emptied into the vehicle's hopper. Once the waste is dumped, it is compacted by a hydraulically powered moving wall that oscillates backwards and forwards to push the waste to the rear of the vehicle. Most of the newer packing trucks have "pack-on-the-go hydraulics" which lets the driver pack loads while driving, allowing faster route times. When the body is full, the compaction wall moves all the way to the rear of the body, ejecting it via an open tailgate. There is also a system called the Curotto Can which is an attachment for a front loader that has an automated arm that functions as an automated side loader that allows the driver to dump carts.

14.5 m^3 rear load container serviced in Copenhagen.

Garbagemen loading garbage by hand in Japan, 2013.

Rear Loaders

Rear loaders have an opening at the rear that a waste collector can throw waste bags or empty the contents of bins into. Often in many areas they have a lifting mechanism to automatically empty large carts without the operator having to lift the waste by hand.

Another popular system for the rear loader is a rear load container specially built to fit a groove in the truck. The truck will have a chain or cable system for upending the container. The waste will then slide into the hopper of the truck.

The modern rear loader usually compacts the waste using a hydraulically powered mechanism that employs a moving plate or shovel to scoop the waste out from the loading hopper and compress it against a moving wall. In most compactor designs, the plate has a pointed edge (hence giving it the industry standard name *packer blade*) which is designed to apply point pressure to the waste to break down bulky items in the hopper before being drawn into the main body of the truck.

Compactor designs, however, have been many and varied, however the two most popular in use today are the "sweep and slide" system (first pioneered on the Leach 2R Packmaster), where the packer blade pivots on a moving carriage which slides back and forth, and the "swing link" system (such as the Dempster Routechief) where the blade literally swings on a "pendulum"-style mechanism. The Heil Colectomatic used a combination of a lifting loading hopper and a pivoting sweeper blade to clear and compact waste in anticipation of the next load.

So-called "continuous" compactors were popular in the 1960s and 1970s. The German Shark design (later Rotopress) used a huge rotating drum, analogous to a cement mixer, in conjunction with a serrated auger to grind down and compact the garbage. SEMAT-Rey of France pioneered the rotating rake system (also used in the British Shelvoke and DrewryRevopak) to both mutilate waste and break down large items. Other systems used a large Archimedes' screw to draw in waste and mutilate it inside the body. A mixture of safety concerns, and higher fuel consumption has seen a decline in the popularity of continuously compacting garbage trucks. The Rotopress design remains popular due to its niche in being able to effectively deal with green waste for composting.

The wall will move towards the front of the vehicle as the pressure forces the hydraulic valves to open, or as the operator moves it with a manual control.

A unique rear-loading system involves a rear loader and a front-loading tractor (usually a Caterpillar front loader with a Tink Claw) for yard waste collection (and in some cities, garbage and recycling). The front loader picks up yard waste set in the street, and then loaded into the back of a rear loader. This system is used in several cities, including San Jose.

Side Loaders

Side loaders are loaded from the side, either manually, or with the assistance of a joystick-controlled robotic arm with a claw, used to automatically lift and tip wheeled bins into the truck's hopper.

Automated Side Loaders

Lift-equipped trucks are referred to as automated side loaders, or ASL's. Similar to a front-end loader, the waste is compacted by an oscillating packer plate at the front of the loading hopper which forces the waste through an aperture into the main body and is therefore compacted towards the rear of the truck. An Automated Side Loader only needs one operator, where a traditional rear load garbage truck may require two or three people, and has the additional advantage of reducing on the job injuries due to repetitive heavy lifting. Due to these advantages, ASL's have become more popular than traditional manual collection. Typically an Automated Side Loader uses standardized wheeled carts compatible with the truck's automated lift.

Semi-automated Side Loaders

An Automated side loader garbage truck in Canberra, Australia.

Semi-automated side loaders use an automated mechanism to lift and dump manually aligned waste containers inside the main body of the truck. The primary difference of semi-automated side loaders is that they require more than one person, to operate the truck, and to manually bring and align containers to the loading hopper on the side of the truck.

Automated garbage collection in Aardenburg, Netherlands.

As with front loaders, the compaction mechanism comprises a metal pusher plate in the collection hopper which oscillates backwards and forwards under hydraulic pressure, pushing the refuse through an aperture, thus compacting it against the material already loaded. On some ASL's there is also a "folding" crusher plate positioned above the opening in the hopper, that folds down to crush bulky items within reach of the metal pusher plate. Another compactor design is the "paddle packer" which uses a paddle that rotates from side to side, forcing refuse into the body of the truck.

Automated side loader in operation on an Autocar truck chassis.

Garbage collection by an automatic side loader during autumn in Kelowna, British Columbia, Canada.

Pneumatic Collection

Volvo pneumatic collector used for "waste suction".

Pneumatic collection trucks have a crane with a tube and a mouthpiece that fits in a hole, usually hidden under a plate under the street. From here it will suck up waste from an underground installation. The system usually allows the driver to "pick up" the waste, even if the access is blocked by cars, snow or other barriers.

Grapple truck.

Grapple Trucks

Grapple trucks enable the collection of bulk waste. A large percentage of items in the solid waste stream are too large or too heavy to be safely lifted by hand into traditional garbage trucks. These items (furniture, large appliances, branches, logs) are called bulky waste or "oversized." The preferred method for collecting these items is with a grapple truck. Grapple trucks have hydraulic knucklebooms, tipped with a clamshell bucket, and usually include a dump body or trailer.

Roll-offs

Roll-offs are characterized by a rectangular footprint, utilizing wheels to facilitate rolling the dumpster in place. The container is designed to be transported by special roll-off trucks. They are relatively efficient for bulk loads of waste.

Sand Cleaning Machine

Raking beach cleaner purifies sand.

A tractor-pulled beach cleaner at Hietaniemi beach in Helsinki, Finland.

A sand cleaning machine, beach cleaner, or (colloquially) sandboni is a vehicle that drags a raking or sifting device over beach sand to remove rubbish and other foreign matter. They are manually self-pulled vehicles on tracks or wheels or pulled by quad-bike or tractor. Seaside cities use beach cleaning machines to combat the problems of litter left by beach patrons and other pollution washed up on their shores. A chief task in beach cleaning strategies is finding the best way to handle waste matter on the beaches, taking into consideration beach erosion and changing terrain. Beach cleaning machines work by collecting sand by way of a scoop or drag mechanism and then raking or sifting anything large enough to be considered foreign matter, including sticks, stones, litter and other items. Similar applications include lake beaches, sandfields for beach volleyball and kindergarten and playing field sandpits. The word "sandboni" is a back-formation referencing the ice-surfacing machine Zamboni.

Common Technologies

Raking technology can be used on dry or wet sand. When using this method, a rotating conveyor belt containing hundreds of tines combs through the sand and removes surface and buried debris while leaving the sand on the beach. Raking machines can remove materials ranging in size from small pebbles, shards of glass, and cigarette butts to larger debris, like seaweed and driftwood. By keeping the sand on the beach and only lifting the debris, raking machines can travel at high speeds.

Sifting technology is practiced on dry sand and soft surfaces. The sand and waste are collected via the pick-up blade of the vehicle onto a vibrating screening belt, which leaves the sand behind. The waste is gathered in a collecting tray which is often situated at the back of the vehicle. Because sand and waste are lifted onto the screening belt, sifters must allow time for the sand to sift through the screen and back onto the beach. The size of the materials removed is governed by the size of the holes in the installed screen.

Combined raking and sifting technology differs from pure sifters in that it uses rotating tines to scoop sand and debris onto a vibrating screen instead of relying simply on the pick-up blade. The tines' position can be adjusted to more effectively guide different-sized materials onto the screen. Once on the screen, combined raking and sifting machines use the same technology as normal sifters to remove unwanted debris from the sand.

Sand sifting by hand is used for smaller areas or sensitive habitat. Sand and debris is collected into a windrow or pile and manually shoveled onto screened sifting trays to separate the debris from the sand. While effective, it requires the movement of sand to the site of the tray, and then redistribution of the sand after sifting. A more efficient method is the use of a screened fork at the place where the debris is located. The effort to manually agitate the sand can become tiresome; however, a recent development of a battery-powered sand rake combines the spot cleaning effectiveness of manual screening with the ease of an auto-sifting hand tool.

Areas of Operation

Sandboni at the gulf of Mexico.

Beach volleyball field after rework of a sand cleaner.

Sand cleaning machines are used all over the world to ensure the safety and happiness of beach-goers. By removing litter, unwanted seaweed, and other debris from the beach, municipalities and resorts are able to maintain their beaches with fewer invested hours.

In addition to their regular litter-removing uses, beach and sand cleaners have been used to clean up after natural disasters. For example:

In Galveston, Texas, low oxygen levels in the water resulted in thousands of dead fish washing ashore. Raking sand cleaners were then used to remove the rotting fish off the beach before they released excessive toxins into the air, sand, and water.

The Olympic Games 2008 saw the first remote-control Sandbonis for the beach volleyball fields in Beijing Chaoyang Park.

The cleanup after the Deepwater Horizon oil spill saw large applications of sand cleaners to the area. Similarly, the Rena oil spill in New Zealand also saw beach cleaners deployed in an effort to remove the affected sand.

Manufacturers

The major manufacturers of large beach-sand cleaning machines are considered to be H Barber & Sons, Cherrington, Beach Tech, Rockland and Tirrenia Srl.

There are many other manufacturers of sand cleaners being used for other purposes. For example, a smaller 4-wheel and halftrack sand cleaning machine is used for sandpits in Kindergarten and municipality playing fields and for beach volleyball. When environmental or spot-cleaning requires hand operations, an auto-sifting, lightweight, screened rake can be the best choice.

Waste Sorting

Recycling bins in Singapore.

Manual waste sorting for recycling.

Characteristic containers for recycling in Portovenere, Italy.

Garbage containers in Fuchū, Tokyo, Japan.

Waste sorting is the process by which waste is separated into different elements. Waste sorting can occur manually at the household and collected through curbside collection schemes, or automatically separated in materials recovery facilities or mechanical biological treatment systems. Hand sorting was the first method used in the history of waste sorting.

Waste Can also be Sorted in a Civic Amenity Site.

Waste segregation means dividing waste into dry and wet. Dry waste includes wood and related products, metals and glass. Wet waste, typically refers to organic waste usually generated by eating establishments and are heavy in weight due to dampness. Waste can also be segregated on basis of biodegradable or non-biodegradable waste.

Landfills are an increasingly pressing problem. Less and less land is available to deposit refuse, but the volume of waste is growing all time. As a result, segregating waste is not just of environmental importance, but of economic concern, too.

Methods

Waste is collected at its source in each area and separated. The way that waste is sorted must reflect local disposal systems. The following categories are common:

- Paper
- Cardboard (including packaging for return to suppliers)
- Glass (clear, tinted – no light bulbs or window panes, which belong with residual waste)
- Plastics
- Textiles
- Wood, leather, rubber
- Scrap metal
- Compost
- Special/hazardous waste
- Residual waste

Organic waste can also be segregated for disposal:

- Leftover food which has had any contact with meat can be collected separately to prevent the spread of bacteria.
    - Meat and bone can be retrieved by bodies responsible for animal waste
    - If other leftovers are sent, for example, to local farmers, they can be sterilised before being fed to the animals
- Peel and scrapings from fruit and vegetables can be composted along with other degradable matter. Other waste can be included for composting, too, such as cut flowers, corks, coffee grindings, rotting fruit, tea bags, egg- and nutshells, paper towels etc.

Chip pan oil (fryer oil), used fats, vegetable oil and the content of fat filters can be collected by companies able to re-use them. Local authority waste departments can provide relevant addresses. This can be achieved by providing recycling bins.

By Country

In Germany, regulations exist that provide mandatory quotas for the waste sorting of packaging waste and recyclable materials such as glass bottles.

In Denpasar, Bali, Indonesia, a pilot project using an automated collecting machine of plastic bottles or aluminium cans with voucher reward has been implemented in a market.

Transfer Station

Transfer station is a centralised facility, where waste is unloaded from smaller collection vehicles and re-loaded into large vehicles for transport to a disposal or processing site. This transfer of waste is frequently accompanied by removal, separation or handling of waste. In areas, where wastes are not already dense, they maybe compacted at a transfer station. The technical limitations of smaller collection vehicles and thelow hauling cost of solid waste, using larger vehicles, make a transfer station viable. Also, the use of transfer station proves reasonable, when there is a need for vehicles servicing a collection route to travel shorter distances, unload and return quickly to their primary task of collecting the waste.

Limitations in hauling solid wastes are the main factors to be considered, while evaluating the use of transfer stations. These include the additional capital costs of purchasing trailers, building transfer stations and the extra time, labour and energy required for transferring wastes from collection truck to transfer trailer.

Consider also the following factors that affect the selection of a transfer station:

- Types of waste received.
- Processes required in recovering material from wastes.
- Required capacity and amount of waste storage desired.
- Types of collection vehicles using the facility.
- Types of transfer vehicles that can be accommodated at the disposal facilities.
- Site topography and access.

The main problem in the establishment of a transfer station, however, is securing a suitable site. Stored solid wastes and recyclable materials, if not properly handled, will attract flies and other insect vectors. Odours from the transferred solid wastes will also be a nuisance, if not properly controlled. In addition, the traffic and noise due to small and large collection vehicles, collectors, drivers, etc., invite the resentment of the communities living in the vicinity of transfer stations (EPA, 1995).

Types

Depending on the size, transfer stations can be either of the following two types:

(i) Small to medium transfer stations: These are direct-discharge stations that provide no intermediate waste storage area. The capacities are generally small (less than 100 tonnes/day) and medium (100 to 500 tonnes/day). Depending on weather, site aesthetics and environmental concerns, transfer operations of this size may be located either indoor or outdoor. More complex small transfer stations are usually attended during hours of operation and may include some simple waste and materials processing facilities. For example, it includes a recyclable material separation and processing centre. The required overall stationcapacity (i.e., the number and size of containers) depends on the size and population density of the area served and the frequency of collection.

(ii) Large transfer stations: These are designed for heavy commercial use by private and municipal collection vehicles. The typical operational procedure for a larger station is as follows:

- when collection vehicles arrive at the site, they are checked in for billing, weighed and directed to the appropriate dumping area;
- collection vehicles travel to the dumping area and empty the wastes into a waiting trailer, a pit or a platform;
- after unloading, the collection vehicle leaves the site, and there is no need to weigh the departing vehicle, if its weight (empty) is known;
- transfer vehicles are weighed either during or after loading. If weighed during loading, trailers can be more consistently loaded to just under maximum legal weights and this maximises payloads and minimises weight violations.

Designs for Larger Transfer Operations

Several different designs for larger transfer operations are common, depending on the transfer distance and vehicle type. Most designs, however, fall into one of the following three categories:

(i) Direct-discharge non-compaction station: In these stations, waste is dumped directly from collection vehicle into waiting transfer trailers and is generally designed with two main operating floors. In the transfer operation, wastes are dumped directly from collection vehicles (on the top floor) through a hopper and into open top trailers on the lower floor. The trailers are often positioned on scales so that dumping can be stopped when the maximum payload is reached. A stationary crane with a bucket is often used to distribute the waste in the trailer. After loading, a cover or tarpaulin is placed over the trailer top. However, some provision for waste storage during peak time or system interruptions should be developed. Because of the use of little hydraulic equipment, a shutdown is unlikely and this station minimises handling of waste.

(ii) Platform/pit non-compaction station: In this arrangement, the collection vehicles dump their wastes onto a platform or into a pit using waste handling equipment, where wastes can be temporarily stored, and if desired, picked through for recyclables or unacceptable materials. The waste is then pushed into open-top trailers, usually by front-end loaders. Like direct discharge stations, platform stations have two levels. If a pit is used, however, the station has three levels. A major advantage of these stations is that they provide temporary storage, which allows peak inflow of wastes to be levelled out over a longer period. Construction costs for this type of facility are usually higher because of the increased floor space. This station provides convenient and efficient storage area and due to simplicity of operation and equipment, the potential for station shutdown is less.

(iii) Compaction station: In this type of station, the mechanical equipment is used to increase the density of wastes before they are transferred. The most common type of compaction station uses a hydraulically powered compactor to compress wastes. Wastes are fed into the compactor through a chute, either directly from collection trucks or after intermediate use of a pit. The hydraulic ram of the compactor pushes waste into the transfer trailer, which is usually mechanically linked to the compactor (EPA, 1995). Compaction stations are used when:

- wastes must be baled for shipment;
- open-top trailers cannot be used because of size restrictions;
- site topography or layout does not accommodate a multi-level building.

The main disadvantage of a compaction facility is that the facility's ability to process wastes is directly dependent on the operative-ness of the compactor. Selection of a quality compactor, regular maintenance of the equipment, easy availability of spare parts and prompt availability of the service personnel are essential for the station's reliable operation.

Capacity

A transfer station should have enough capacity to manage and handle the wastes at the facility throughout its operating life. While selecting the design capacity of a transfer station, we must, therefore, consider trade-offs between the capital costs associated with the station and equipment and the operational costs. Designers should also plan adequate space for waste storage and, if necessary, waste processing. Transfer stations are usually designed to have 1.5– 2 days of storage capacity. The collection vehicle unloading area is usually the waste storage area and sometimes a waste sorting area. When planning the unloading area, designers should allow adequate space for vehicle and equipment manoeuvring. To minimise the space required, the facility should be designed such that the collection vehicle backs into the unloading position. Adequate space should also be available for offices, employee facilities, and other facility-related activities (EPA, 1995). Factors that should be considered in determining the appropriate capacity of a transfer facility include:

- capacity of collection vehicles using the facility;
- desired number of days of storage space on tipping floor; time required to unload collection vehicles;
- number of vehicles that will use the station and their expected days and hours of arrival;
- waste sorting or processing to be accomplished at the facility; transfer trailer capacity;
- hours of station operation;
- availability of transfer trailers waiting for loading;
- time required, if necessary, to attach and disconnect trailers from tractors or compactors.

Transfer station capacity can be determined using the following formulae:

(i) Pit stations: Based on the rate at which wastes can be unloaded from collection vehicles:

$$C = P_c \text{ x } (L/W) \text{ x } (60 \text{ x } H_w/T_c) \text{ x } F$$

Based on rate at which transfer trailers are loaded:

$$C = (P_t \text{ x } N \text{ x } 60 \text{ x } H_t) / (T_t + B)$$

(ii) Direct dump stations:

$$C = (Nn \text{ x } Pt \text{ x } F \text{ x } 60 \text{ x } Hw) / [((Pt/Pc) \text{ x } (W/Ln)) \text{ x } Tc + B]$$

(iii) Hopper compaction stations:

$$C = (N_n \text{ x } P_t \text{ x } F \text{ x } 60 \text{ x } H_w) / [(P_t/P_c \text{ x } T_c) + B]$$

(iv) Push pit compaction station:

$$C = (N_p \text{ x } P_t \text{ x } F \text{ x } 60 \text{ x } H_w) / [(P_t/P_c \text{ x } W/L_p \text{ x } T_c) + B_c + B]$$

Where:

C = Station capacity (tonnes/day) ; P_c= Collection vehicle payload (tonnes) ; L = Total length of dumping space (feet) ; H_w = Hours per day that waste is delivered; T_c = Time (in minutes) to unload each collection vehicle; F = Peaking factor (ratio of the number of collection vehicles received during an average 30-minute period to the number received during a peak 30-minute period) ; L_p = Length of push pit (feet) ; N_p = Number of push pits; B_c = Total cycle time for clearing each push pit and compacting waste into trailer; P_t = Transfer trailer payload (tonnes) ; N = Number of transfer trailers loading simultaneously; H_t = Hours per day used to load trailers (minutes) ; B = Time to remove and replace each loaded trailer (minutes) ; T_t = Time to load each transfer trailer (minutes) ; N_n = Number of hoppers; L_n = Length of each hopper (feet).

These formulae are useful in estimating the capacity of various types of transfer stations (EPA, 1995) and should be adapted, as necessary, for specific applications.

Viability

Transfer stations offer benefits such as lower collection costs (because crews waste less time travelling to the site), reduced fuel and maintenance costs for collection vehicles, increased flexibility in selection of disposalfacilities, opportunity to recover recyclables or compostables at the transfer site and the opportunity to shred or scoop wastes prior to disposal. These benefits must be weighed against the costs to develop and operate the facility.

The classical approach to arrive at the economic viability of operating a transfer station, is to add the unit cost of the transfer station to the cost of hauling using large vehicles, and to compare this cost with the cost of hauling directly to the disposal site using the smaller vehicles that service the collection area. The cost of hauling using small vehicles is the sum of the depreciation cost of the vehicle, driver's salary, salary of the collection crew (if they are on standby waiting for the vehicle to return to the collection area) and fuel cost. The transfer station cost is the sum of the transfer station's depreciation cost and the operating and maintenance costs divided by the capacity of the station. The cost of using the large vehicle is the sum of the vehicle depreciation, fuel cost and driver's salary.

The cost-effectiveness of a transfer station depends on the distance of disposal site from thegeneration area, andadistance of 10 – 15 km is usually the minimum cost-effective distance (Phelps, et al., 1995). The distance between the disposal site and collection area is one of the principal variables in deciding whether to use a transfer station or haul the solid wastes directly from the collection area to the disposal site. Figure illustrates the economic analysis involving the effect of the hauling distance on the collection cost:

Cost Analysis to Determine Viability of Transfer Station

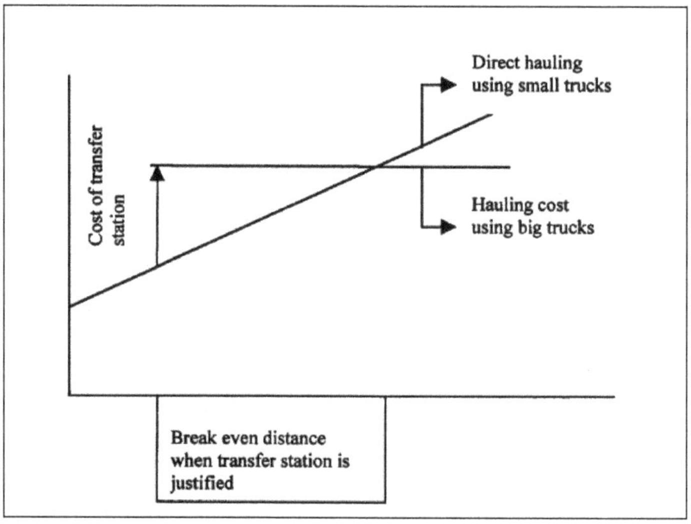

Now, let us consider first the case in which the transfer station is located direc tly along the hauling route between the disposal site and the collection area. Let the unit cost of hauling using a small vehicle be Rs. A/m3 km. The cost of operation, maintenance, depreciation, loading and unloading at the transfer station be Rs. B/m3 and the cost of hauling using large vehicles beRs.C/m3 km.Ifthe distance between the collection area and the transfer station is X km and the distance between the transfer station and the disposal site is Y km, then the distance between the collection area and the disposal site is X + Y km. Then, the total cost of hauling the solid wastes from the collection area to the disposal site using a transfer station is:

$$T = 2AX + B + 2CY$$

The factor 2 is added to account for the round trip, which effectively doubles the distance travelled. The total cost of hauling without the transfer station is:

$$T1 = 2A(X + Y)$$

The transfer station is justified, when:

$$T < T_1$$

That is, the hauling cost using a transfer station is lower than the direct hauling costs between the collection area and the disposal site. Substituting the values of T and T_1 yields:

$$2AX + B + 2CY < 2AX + 2AY$$

or

$$Y > B/(2A - 2C)$$

Note that X cancels out. The distance between the potential transfer station site and the disposal site is the variable to consider. The distance between the collection area and the disposal site is important in deciding the utilisation of a transfer station, if X is equal to zero, in which case the transfer station is located right at the centroid of the collection area. Under normal conditions, the centroid of the collection area has a high land value, and it would be impractical to locate a solid waste transfer station in this area. Figure shows the effect of the distance between the potential transfer station site and the disposal site on the hauling cost.

Consider a general case in which the transfer station is located away from the hauling route between the collection area and the disposal site. Let Z be the additional distance travelled by the vehicles. The cost T, when using a transfer station, is then equal to:

$$T = B + 2AX + 2AZ + 2CY + 2CZ$$

The cost of direct hauling from the collection area to the disposal site remains the same as previously defined.The use of a transfer station is justified, if:

$$B + 2AX + 2AZ + 2CY + 2AZ < 2AX + 2AY$$

or

$$Y > (B + 2CZ + 2AZ) /(2A - 2C)$$

Record Keeping, Control, Inventory and Monitoring

For effective waste collection and, indeed, SWM, we must maintain records on the quantities of wastes collected and their variation within a week, month and year, as well as on established long-term trends in solid waste generation ratesand composition, sources of wastes and the personnel collecting them. Long- term trends in solid waste generation rates and composition form the basis for planning, especially in budgeting for future vehicle requirements, allocating the collection vehicles and crew, building transfer stations, acquiring strategic lands and determining disposal options. Table contains an illustration of a checklist of factors that affect the waste collection system:

Checklist of Variables Affecting Collection System,

Components	Factors to Consider
Crew size	labour cost distance between containers size and types of containers loading accessories available in the truck collection vehicle used
Container type	solid wastes generation rate density of waste generation street width traffic vol ume collection crew configuration standard of li ving

Collection accessory	labour cost protection of worker's health
Vehicle size/type	streetwidth, trafficvolumesolidwastegenerationratescrewsize viabilityofatrans ferstation
Collection route	street width, traffic volume solid waste generation rates crew size viability of a transfer station street width, traffic volume direction of traffic flow solid waste generation rates spatial distribution of wastes local topography
Trans fer station	distance between disposal site and collection area hauling cost for small and large trucks cost of transferring the solid wastes from small to large trucks

Records of personnel and quantities of wastes collected are, when maintained, useful in determining the efficiency of the personnel and in correlating waste quantities with conditions in the service area. A time keeping system at the transfer or disposal site is a key element in improving the efficiency of collection system and planning an upgraded system. The timekeeping system determines if the crewwere taking long rest periods, spending time salvaging or carrying out unauthorised activities. The performance of a particular crew in terms of the quantity of solid wastes collected per day could be compared with that of another collection crew working under similar conditions.

The composition of solid wastes should be measured at least once a year for major districts and possibly once every two years in residential areas with stagnant growth rates and development. Changes in composition affect the collection equipment and configuration of the collection system is important in designing the disposal system. Changes in an energy source (such as a shift to gas or electricity from wood or charcoal for cooking and heating), reduces the ash content of wastes, making the solid waste lighter, in which case, larger containers could be used. The same line of analysis holds true in specifying the collection vehicles. Comparison of the routes taken by various crew serving a particular area helps to identify the best hauling route. Although this route may be longer, it could be more economical in terms of hauling time. However, note that the best route often changes with the season.

All these decisions should be based on reliable data, without which the waste collection system will inevitably be ineffective. Proper interpretation of monitoring data allows the authority to adapt the proposed system to actual conditions. In some instances, it also allows management to identify areas, where the design is not realistic.

Implementing Collection and Transfer System

Implementing of collection and transfer system involves the following activities, which are important for success of the plan (EPA, 1995):

(i) Finalising and implementing the system management plan: For proper implementation of collection and transfer system, it is necessary to have clear organisational structures and management plans. The organisational structure shouldbe simple, with a minimumofadministrative andmanagement layers between collection crews and top management. All workers in the department should clearly understand the department's mission and their roles. Through training, incentives and reinforcement by management, workers should be encouraged to be customer-oriented and team contributors. Feedback mechanisms must be introduced to help the crew review their performance and help managers monitoring the performance of crews, equipment, etc. It is also important to periodically review the management plans and structures, as implementation of collection services continues.

(ii) Purchasing and managing equipment: For purchasing equipment, most municipalities issue bid specifications. Detailed specifications include exact requirements for equipment sizes and capacities, power ratings, etc. Performance specifications often request that equipment be equivalent to certain available models and meet standards for capacity, speed, etc. Municipalities may either perform equipment maintenance themselves, contract with a local garage, or in some cases, contract with the vehicle vendor at the time of purchase. As part of the preventive maintenance programme, the collection crew should check the vehicle chassis, tyres and body daily and report any problems to maintenance managers. In addition, each vehicle should have an individual maintenance record that includes the following items:

- preventive maintenance schedule;
- current list of specific engine;
- a description of repairs and a list containing information on the repair date, mechanic, cost, type and manufacturer of repair parts and the length of time the truck was out of service, for each maintenance event.

(iii) Hiring and training personnel: As in all organisations, good personnel management is essential to an efficient, high-quality waste collection system. Authorities responsible for SWM should, therefore, strive to hire and keep well-qualified personnel. The recruitment programme shouldassess applicants' abilities to perform the types of physical labour required for the collection, equipment and methods used. To retain employees, management should provide a safe working environment that emphasises career advancement, participatory problem solving and worker incentives. Worker incentives should be developed to recognise and reward outstanding performance by employees. Ways to accomplish motivation include merit-based compensation, awards programme and a work structure. Feedback on employee performance should be regular and frequent.

Safety is especially important because waste collection employees encounter many hazards during each workday. As a result of poor safety records, insurance costs for many collection services are high. To minimise injuries, haulers should have an ongoing safety programme. This programme should outline safety procedures and ensure that all personnel are properly trained on safety issues. Haulers should develop an employee- training programme that helps employees improve and broaden the range of their job-related skills. Education should address such subjects as driving skills, first aid, safe lifting methods, identification of household hazardous wastes, avoidance of substance abuse and stress management.

(iv) Providing public information: Maintaining good communication with the public is important to a well-run collection system. Residents can greatly influence the performance of the collection system by co-operating in separation requirements, and by keeping undesirable materials from entering the collected waste stream. Commonly used methods of communicating information include brochures, articles in community newsletters, newspaper articles, announcements, and advertisements on radio and television, information attachments to utility bills (either printed or given separately) and school handouts. Communication materials should be used to help residents understand the community waste management challenges and the progress in meeting them. Residents should also be kept informed about issues such as the availability and costs of landfillcapacity so that they develop an understanding of the issues and a desire to help meet their waste management needs.

(v) Monitoring system cost and performance: Collection and transfer facilities should develop and maintain an effective system for cost and performance reporting. Each collection crew should complete a daily report containing the following information:

- Total quantity hauled.
- Total distance and travel times to and from the disposal site.
- Amounts delivered to each disposal, transfer, or processing facility.
- Waiting time at sites.
- Number of loads hauled.
- Vehicle or operational problems needing attention.

Collected data should be used to forecast workloads, truck costs, identify changes in the generation of wastes and recyclables, trace the origin of problematic waste materials and evaluate crew performance. Just as the goals of a collection programme set its overall directions, a monitoring system provides the short-term feedback necessary to identify the corrections needed to achieve those goals.

In brief guidelines for planning waste collection and transport are given below:

- Analyse the quantum of waste generated with composition.
- Capacity building of town municipalities with appropriate infrastructure and the knowledge of existing laws or regulations on waste collection, transport and safe disposal.
- Designate a para-state agency to oversee waste collection, transport and disposal to avoid confusion among para-state government agencies.
- Determine geographic scope of collection and transport services. Determine funding, equipment and labour needs.
- Determine the type and amount of waste to be processed.
- Implement decentralised waste treatment through proven local techniques.
- Deploy GPS (Global Positioning System) based trucks for waste collection and transport to minimise pilferages.
- Adopt spatial information system for the management.

- Consider a transfer station that serves as a central location for activities to sort and recover waste.
- Implement decentralised waste management including all stakeholders with active participation of the public.

References

- "Memorandum submitted by Essex Friends of the Earth". www.publications.parliament.uk. Retrieved 11 January 2020
- Geroux, Zachary; Voytko, Eric. "The Ever Expanding History of the Front Load Refuse Truck". Retrieved 10, June 2020
- "The State of New Zealand's Environment". Ministry for the Environment (New Zealand). 1997. Retrieved 01, July 2020
- Motor-Cars for Dust Collection", The Automotor and Horseless Carriage Journal, February 1897, p192
- Ministry for the Environment (December 2007). Environment New Zealand 2007. Ministry for the Environment (New Zealand). ISBN 978-0-478-30192-2

Chapter 4

Waste Disposal

Waste disposal refers to the various activities and actions necessary for the management of waste from its origin to final disposal. Some of the practices elaborated in this chapter related to waste disposal are disposal options and selection criteria, landfill, sanitary landfill, landfill gas emission, leachate formation, landfill operation issues, etc. This chapter discusses in detail the theories related to waste disposal.

Disposal Options and Selection Criteria

The most common disposal option practised currently in many countries is either uncontrolled dumping or dumping with moderate control. The environmental costs of uncontrolled dumping include breeding of disease causing vectors (e.g., flies, mosquitoes and rodents), pollution, odour and smoke.

Disposal Options

In this Section, the options available for waste disposal is discussed, and in that respect, we will consider the following:

(i) Uncontrolled dumping or non-engineered disposal: As mentioned, this is the most common method being practised in many parts of the world, and India is no exception. In this method, wastes are dumped at a designated site without any environmental control. They tend to remain there for a long period of time, pose health risks and cause environmental degradation. Due to the adverse health and environmental impact associated with it, the non-engineered disposal is not considered a viable and safe option.

(ii) Sanitary landfill: Unlike the non-engineered disposal, sanitary landfill is a fully engineered disposal option in that the selected location or wasteland is carefully engineered in advance before it is pressed into service. Operators of sanitary landfills can minimise the effects of leachate (i.e., polluted water which flows from a landfill) and gas production through proper site selection, preparation and management. This particular option of waste disposal is suitable when the land is available at an affordable price, and adequate workforce and technical resources are available to operate and manage the site.

(iii) Composting: This is a biological process of decomposition in which organisms, under controlled conditions of ventilation, temperature and moisture, convert the organic portion of solid waste into humus-like material. If this process is carried out effectively, what we get as the final product is a stable, odour-free soil conditioner. Generally, the option of composting is considered, when a considerable amount of biodegradable waste is available in the waste stream and there is use or market for composts. Composting can be either centralised orsmall-scale. Centralised composting plants are possible, if adequate skilled workforce and equipments are available. And, small-scale composting practices can be effective at household level, but this needs public awareness.

(iv) Incineration: This refers to the controlled burning of wastes, at a high temperature (roughly 1200 – 1500 C), which sterilises and stabilises the waste in addition to reducing its volume. In the process, most of the combustible materials(i.e.,self-sustainingcombustible matter, whichsavestheenergyneeded tomaintain the combustion) such as paper or plastics get converted into carbon dioxide and ash. Incineration may be used as a disposal option, when land filling is not possible and the waste composition is highly combustible. An appropriate technology, infrastructure and skilled workforce are required to operate and maintain the plant.

(v) Gasification: This is the partial combustion of carbonaceous material (through combustion) at high temperature (roughly 1000 C) forming a gas, comprising mainly carbon dioxide, carbon monoxide, nitrogen, hydrogen, water vapour and methane, which can be used as fuel.

(vi) Refuse-derived fuel (RDF): This is the combustible part of raw waste, separated for burning as fuel. Various physical processes such as screening, size reduction, magnetic separation, etc., are used to separate the combustibles.

(vii) Pyrolysis: This is the thermal degradation of carbonaceous material to gaseous, liquid and solid fraction in the absence of oxygen. This occurs at a temperature between 200 and 900 C.The product of pyrolysis is a gas ofrelatively high calorific value of 20,000 joules per gram with oils, tars and solid burned residue (Ali, et al 1999).

Relative Merits of Some Options

Having touched upon several disposal options, let us now present the merits and demerits of some of them in Table:

Relative Merits of Disposal Options

Disposal Option / Sustainability Indicator	Non- engineered Disposal	Sanitary Landfill	Composting	Incineration
Volume reduction	×	×	×	√
Expensive	×	√	√	√
Long term maintenance	√	√	×	×
By product recovery	×	√	√	√
Adaptability to all wastes	√	√	×	×
Adverse environmental effect	√	√	×	√

Selection Criteria

With the help of proper frameworks and sub-frameworks, we can assess the effectiveness of each of the waste disposal options. While a framework represents an aid to decision-making and helps to ensure the key issues are considered, a sub-framework explains how and why the necessary information should be obtained (Ali, et al 1999). A framework contains a list of issues and questions pertaining to the technical, institutional, financial, social and environmental features of a waste disposal system to assess the capacity of a disposal option to meet the requirements. For example, an appraisal of waste disposal option must include the following:

(i) Technical: This feature, involving efficient and effective operation of the technology being used, evaluates the following components of a SWM system:

- composition of wastes, e.g., type, characteristics and quantity.
- existing practices, e.g., collection, transport, and recycling process.
- siting, e.g., location of disposal site, engineering material, etc.
- technology, e.g., operation, maintenance, technical support, etc.
- impact, e.g., anticipated by-product, requirement for their treatment and disposal, etc.

(ii) Institutional: This involves the ability and willingness of responsible agencies to operate and manage the system by evaluating the following:

- structures,roles and responsibilities, e.g., currentinstitutional frameworks.
- operational capacity, e.g., municipal capacities, local experience and staff training.
- incentives, e.g.,management improvementandwastedisposal practices.
- innovation and partnership.

(iii) Financial: This assesses the ability to finance the implementation, operation and maintenance of the system by evaluating the following:

- financing and cost recovery, e.g., willingness to raise finance for waste management.
- current revenue and expenditure on waste management.
- potential need for external finance for capital cost.

(iv) Social: This helps in avoiding adverse social impact by evaluating the following:

- waste picking, which has an impact on livelihood and access to waste pickers.
- health and income implication.
- public opinions on the existing and proposed system.

(v) Environmental: This means setting up an environment friendly disposal system by evaluating the following:

- initial environmental risks, i.e., impact of existing and proposed disposal option.
- long-term environmental risks, i.e., long-term implication (future impacts).

Landfill

A landfill in Poland.

A landfill site (also known as a tip, dump, rubbish dump, garbage dump or dumping ground and historically as a midden) is a site for the disposal of waste materials by burial and the oldest form of waste treatment (although the burial part is modern; historically, refuse was just left in piles or thrown into pits). Historically, landfills have been the most common method of organized waste disposal and remain so in many places around the world.

Some landfills are also used for waste management purposes, such as the temporary storage, consolidation and transfer, or processing of waste material (sorting, treatment, or recycling).

A landfill also may refer to ground that has been filled in with rocks instead of waste materials, so that it can be used for a specific purpose, such as for building houses. Unless they are stabilized, these areas may experience severe shaking or soil liquefaction of the ground during a large earthquake.

Operations

Typically, operators of well-run landfills for non-hazardous waste meet predefined specifications by applying techniques to:

1. confine waste to as small an area as possible
2. compact waste to reduce volume
3. cover waste (usually daily) with layers of soil

One of several landfills used by Dryden, Ontario, Canada.

During landfill operations a scale or weighbridge may weigh waste-collection vehicles on arrival and personnel may inspect loads for wastes that do not accord with the landfill's waste-acceptance criteria. Afterward, the waste-collection vehicles use the existing road network on their way to the tipping face or working front, where they unload their contents. After loads are deposited, compactors or bulldozers can spread and compact the waste on the working face. Before leaving the landfill boundaries, the waste collection vehicles may pass through a wheel-cleaning facility. If necessary, they return to the weighbridge for re-weighing without their load. The weighing process can assemble statistics on the daily incoming waste-tonnage, which databases can retain for record keeping. In addition to trucks, some landfills may have equipment to handle railroad containers. The use of "rail-haul" permits landfills to be located at more remote sites, without the problems associated with many truck trips.

Typically, in the working face, the compacted waste is covered with soil or alternative materials daily. Alternative waste-cover materials include chipped wood or other "green waste", several sprayed-on foam products, chemically "fixed" bio-solids, and temporary blankets. Blankets can be lifted into place at night and then removed the following day prior to waste placement. The space that is occupied daily by the compacted waste and the cover material is called a daily cell. Waste compaction is critical to extending the life of the landfill. Factors such as waste compressibility, waste-layer thickness and the number of passes of the compactor over the waste affect the waste densities.

Advantages

Landfills are often the most cost-efficient way to dispose of waste, especially in countries like the United States with large open spaces. While resource recovery and incineration both require extensive investments in infrastructure, and material recovery also requires extensive manpower to maintain, landfills have fewer fixed—or ongoing—costs, allowing them to compete favorably. In addition, landfill gas can be upgraded to natural gas—landfill gas utilization—which is a potential revenue stream. Another advantage is having a specific location for disposal that can be monitored, where waste can be processed to remove all recyclable materials before tipping.

Social and Environmental Impact

Landfill operation in Hawaii. Note that the area being filled is a single, well-defined "cell" and that a protective landfill liner is in place (exposed on the left) to prevent contamination by leachates migrating downward through the underlying geological formation.

Landfills have the potential to cause a number of issues. Infrastructure disruption, such as damage to access roads by heavy vehicles, may occur. Pollution of local roads and water courses from wheels on vehicles when they leave the landfill can be significant and can be mitigated by wheel washing systems. Pollution of the local environment, such as contamination of groundwater or aquifers or soil contamination may occur, as well.

Leachate

Extensive efforts are made to capture and treat leachate from landfills before it reaches groundwater aquifers, but engineered liners always have a lifespan, though it may be 100 years or more. Eventually, every landfill liner will leak, allowing the leachate to contaminate the groundwater. Installation of composite liners with flexible membrane and soil barrier is enforced by the EPA to ensure that leachate is withheld.

Dangerous Gases

Rotting food and other decaying organic waste allows methane and carbon dioxide to seep out of the ground and up into the air. Methane is a potent greenhouse gas, and can itself be a danger because it is flammable and potentially explosive. In properly managed landfills, gas is collected and utilized. This could range from simple flaring to landfill gas utilization. Carbon dioxide is the most widely produced greenhouse gas. It traps heat in the atmosphere, contributing to climate change.

Infections

Poorly run landfills may become nuisances because of vectors such as rats and flies which can cause infectious diseases. The occurrence of such vectors can be mitigated through the use of daily cover.

Other potential issues include wildlife disruption, dust, odor, noise pollution, and reduced local property values.

Landfill Gas

Gases are produced in landfills due to the anaerobic digestion by microbes. In a properly managed landfill this gas is collected and used. Its uses range from simple flaring to the landfill gas utilization and generation of electricity. Landfill gas monitoring alerts workers to the presence of a build-up of gases to a harmful level. In some countries, landfill gas recovery is extensive; in the United States, for example, more than 850 landfills have active landfill gas recovery systems.

Regional Practice

A landfill in Perth, Western Australia.

South East New Territories Landfill, Hong Kong.

Canada

Landfills in Canada are regulated by provincial environmental agencies and environmental protection acts (EPA). Older facilities tend to fall under current standards and are monitored for leaching. Some former locations have been converted to parkland.

European Union

In the European Union, individual states are obliged to enact legislation to comply with the requirements and obligations of the European Landfill Directive. In the UK this is the Waste Implementation Programme.

United Kingdom

Landfilling practices in the UK have had to change in recent years to meet the challenges of the European Landfill Directive. The UK now imposes landfill tax upon biodegradable waste which is put into landfills. In addition to this the Landfill Allowance Trading Scheme has been established for local authorities to trade landfill quotas in England. A different system operates in Wales where authorities are not able to 'trade' between themselves, but have allowances known as the Landfill Allowance Scheme.

United States

U.S. landfills are regulated by each state's environmental agency, which establishes minimum guidelines; however, none of these standards may fall below those set by the United States Environmental Protection Agency (EPA).

Permitting a landfill generally takes between 5 and 7 years, costs millions of dollars and requires rigorous siting, engineering and environmental studies and demonstrations to ensure local environmental and safety concerns are satisfied.

Microbial Topics

The status of a landfill's microbial community may determine its digestive efficiency.

Bacteria that digest plastic have been found in landfills.

Reclaiming Materials

Landfills can be regarded as a viable and abundant source of materials and energy. In the developing world, waste pickers often scavenge for still-usable materials. In a commercial context, landfill sites have also been discovered by companies, and many have begun harvesting materials and energy . Well known examples are gas recovery facilities. Other commercial facilities include waste incinerators which have built-in material recovery. This material recovery is possible through the use of filters (electro filter, active carbon and potassium filter, quench, HCl-washer, SO_{2-}washer, bottom ash-grating, etc.).

Alternatives

In addition to waste reduction and recycling strategies, there are various alternatives to landfills, including waste-to-energy incineration, anaerobic digestion, composting, mechanical biological treatment, pyrolysis and plasma arc gasification, which have all begun to establish themselves in the market. Depending on local economics and incentives, these can be made more financially attractive than landfills.

Restrictions

Countries including Germany, Austria, Sweden, Denmark, Belgium, the Netherlands, and Switzerland, have banned the disposal of untreated waste in landfills. In these countries, only the ashes from incineration or the stabilized output of mechanical biological treatment plants may still be deposited.

Sanitary Landfill

The term landfill generally refers to an engineered deposit of wastes either in pits/trenches or on the surface. And, a sanitary landfill is essentially a landfill, where proper mechanisms are available to control the environmental risks associated with the disposal of wastes and to make available the land, subsequent to disposal, for other purposes. However, you mustnotethata landfill need not necessarily be an engineered site, when the waste is largely inert at final disposal, as in rural areas, where wastes contain a large proportion of soil and dirt. This practice is generally designated as non-engineered disposal method. When compared to uncontrolled dumping, engineered landfills are more likely to have pre-planned installations, environmental monitoring, and organised and trained workforce. Sanitary landfill implementation, therefore, requires careful site selection, preparation and management.

The four minimum requirements you need to consider for a sanitary landfill are:

(i) full or partial hydrological isolation;

(ii) formal engineering preparation;

(iii) permanent control;

(iv) planned waste emplacement and covering.

Against this background, let us now discuss the principles, processes and operation of sanitary landfills.

Principle

The purpose of land filling is to bury or alter the chemical composition of the wastes so that they do not pose any threat to the environment or public health. Landfills are not homogeneous and are usually made up of cells in which a discrete volume of waste is kept isolated from adjacent waste cells by a suitable barrier. The barriersbetween cells generally consist of a layer of natural soil (i.e., clay), which restricts downward or lateral escape of the waste constituents or leachate.

Land filling relies on containment rather than treatment (for control) of wastes. If properly executed, it is a safer and cheaper method than incineration. An environmentally sound sanitary landfill comprises appropriate liners for protection of the groundwater (from contaminated leachate), run-off controls, leachate collection and treatment, monitoring wells and appropriate final cover design (Phelps, 1995).

Figure gives a schematic layout of sanitary landfill along with its various components:

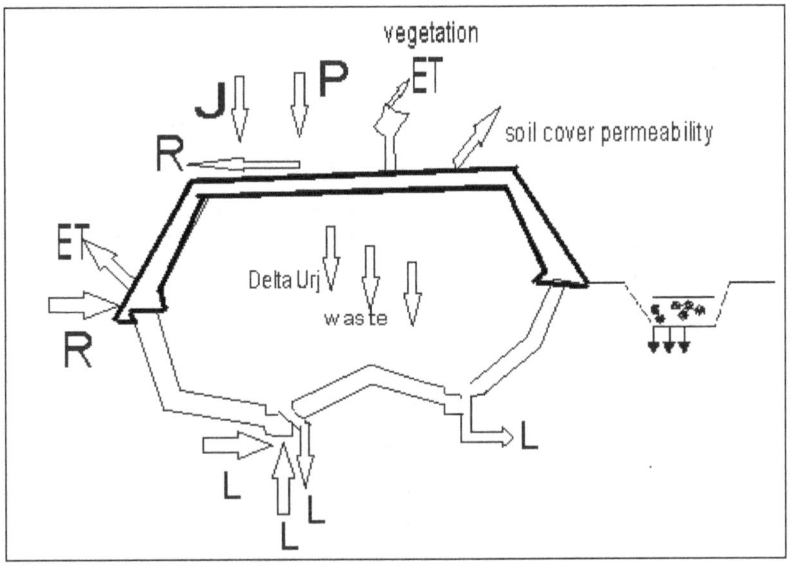

Schematic Layout of Sanitary Landfill.

- Planning phase: This typically involves preliminary hydro-geological and geo-technical site investigations as a basis for actual design.
- Construction phase: This involves earthworks, road and facility construction and preparation (liners and drains) of the fill area.
- Operation phase (5 – 20 years): This phase has a high intensity of traffic, work at the front of the fill, operation of environmental installations and completion of finished sections.
- Completed phase (20 – 100 years): This phase involves the termination of the actual filling to the time when the environmental installations need no longer be operated. The emissions may have by then decreased to a level where they do not need any further treatment and can be discharged freely into the surroundings.
- Final storage phase: In this phase, the landfill is integrated into the surroundings for other purposes, and no longer needs special attention.

Landfill Processes

The feasibility of land disposal of solid wastes depends on factors such as the type, quantity and characteristics of wastes, the prevailing laws and regulations, and soil and site characteristics. Let us now explain some of these processes.

(i) Site selection process and considerations: This requires the development of a working plan – a plan, or a series of plans, outlining the development and descriptions of site location, operation, engineering and site restoration. Considerations for site include public opinion, traffic patterns and congestion, climate, zoning requirements, availability of cover material and liner as well, high trees or buffer in the site perimeter, historic buildings, and endangered species, wetlands, and site land environmental factors, speed limits, underpass limitations, load limits on roadways, bridge capacities, and proximity of major roadways, haul distance, hydrology and detours.

(ii) Settling process: The waste body of a landfill undergoes different stagesof settling or deformation. Figure below illustrates these stages:

Figure Settling Processes in Landfill.

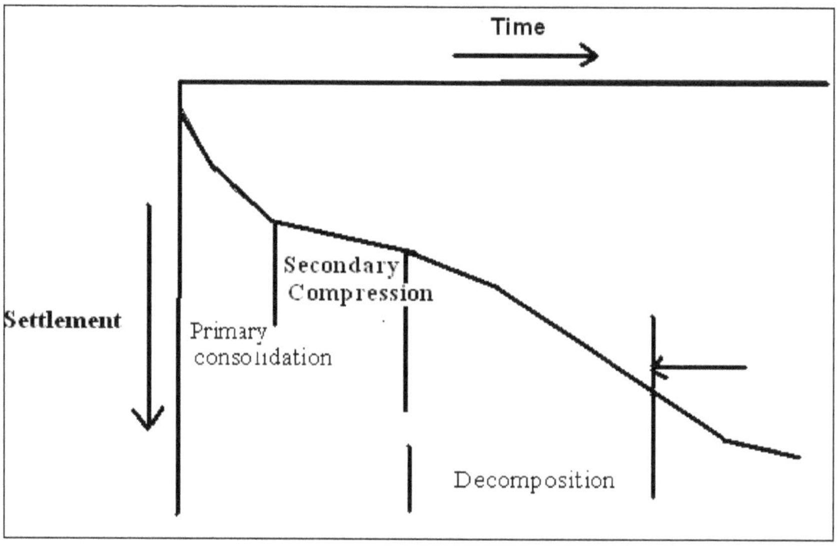

The three stages shown in the figure above are described below:

- Primary consolidation: During this stage, a substantial amount of settling occurs. This settlement is caused by the weight of the waste layers. The movement of trucks, bulldozers or mechanical compactors will also enhance this process. After this primary consolidation, or short-term deformation stage, aerobic degradation processes occur.
- Secondary compression: During this stage, the rate of settling is much lower than that in the primary consolidation stage, as the settling occurs through compression, which cannot be enhanced.
- Decomposition: During the degradation processes, organic material is converted into gas and leachate. The settling rate during this stage increases compared to the secondary compression stage, and continues until all decomposable organic matter is degraded. The settling rate, however, gradually decreases with the passage of time.

To appropriately design protective liners, and gas and leachate collection systems, it is, therefore, necessary to have a proper knowledge of the settling process of wastes.

(iii) Microbial degradation process: The microbial degradation process is the most important biological process occurring in a landfill. These processes induce changes in the chemical and physical environment within the waste body, which determine the quality of leachate and both thequality andquantity of landfill gas. Assuming that landfills mostly receive organic wastes, microbial processes will dominate the stabilisation of the waste and therefore govern landfill gas generation and leachate composition. Soon after disposal, the predominant part of the wastes becomes anaerobic, and the bacteria will start degrading the solid organic carbon, eventually toproduce carbon dioxide and methane. The anaerobic degradation process undergoes the following stages:

- Solid and complex dissolved organic compounds are hydrolysed and fermented by the fermenters primarily to volatile fatty acids, alcohols, hydrogen and carbon dioxide.
- An acidogenic group of bacteria converts the products of the first stage to acetic acid, hydrogen and carbon dioxide.
- Methanogenic bacteria convert acetic acid to methane and carbon dioxide and hydrogenophilic bacteria convert hydrogen and carbon dioxide to methane.

The biotic factors that affect methane formation in the landfill are pH, alkalinity, nutrients, temperature, oxygen and moisture content.

Enhancement of Degradation

Enhancement of the degradation processes in landfills will result in a faster stabilisation of the waste in the landfill, which enhances gas production, and we can achieve this by:

- Adding partly composted waste: As the readily degradable organic matter has already been decomposed aerobically, the rapid acid production phase is overcome, and the balance of acidand methane production bacteria can develop earlier and the consequent dilution effect lowers the organic acid concentration.
- Recirculating leachate: This may have positive effects since a slow increase in moisture will cause a long period of gas production. During warmer periods, recirculated leachate will evaporate, resulting in lower amounts of excess leachate.

Landfill Gas and Leachate

Leachate and landfill gas comprise the major hazards associated with a landfill. While leachate may contaminate the surrounding land and water, landfill gas can be toxic and lead to global warming and explosion leading to human catastrophe (Phelps, 1995). (Note that global warming, also known as greenhouse effect, refers to the warming of the earth's atmosphere by the accumulation of gases (e.g., methane, carbon dioxide and chlorofluorocarbons) that absorbs reflected solar radiation.) The factors, which affect the production of leachate and landfill gas, are the following:

- Nature of waste: The deposition of waste containing biodegradable matter invariably leads to the production of gas and leachate, and the amount depends on the content of biodegradable material in the waste.
- Moisture content: Most micro-organisms require a minimum of approximately 12% (by weight) moisture for growth, and thus the moisture content of landfill waste is an important factor in determining the amount and extent of leachate and gas production.
- pH: The methanogenic bacteria within a landfill produce methane gas, which will grow only at low pH range around neutrality.
- Particle size and density: The size of waste particle affects the density that can be achieved upon compaction and affects the surface area andhence volume. Both affect moisture absorption and therefore are potential for biological degradation.

- Temperature: An increase in temperature tends to increase gas production. The temperature affects the microbial activity to the extent that it is possible to segregate bacteria, according to their optimum temperature operating conditions.

Note that the composition of waste, which varies with region and climate (season), determines the variation in pollution potential. Carbohydrates comprise a large percentage of biodegradable matter within municipal waste, the overall breakdown of which can be represented by the following equation:

(Bacteria)

$$C_6H_{12}O6 \rightarrow CH_4 \quad + \quad CO_2 \quad + \quad Biomass + Heat$$

(Carbohydrate) (Methane) (Carbon dioxide)

Landfill Gas Emission

Landfill gas contains a high percentage of methane due to the anaerobic decomposition of organic matter, which can be utilised as a source of energy.

Composition and Properties

We can predict the amount and composition of the gas generated for different substrates, depending on the general anaerobic decomposition of wastes added. Climatic and environmental conditions also influence gas composition. Due to the heterogeneous nature of the landfill, some acid-phase anaerobic decomposition occurs along with the methanogenic decomposition.Since aerobic and acid-phase degradation give rise to carbon dioxide and not methane, there may be a higher carbon dioxide content in the gas generated than what would otherwise be expected. Furthermore, depending on the moisture distribution, some carbon dioxide goes into solution. This may appearto increase (artificially) the methane content of the gas measured in the landfill. A typical landfill gas contains a number of components such as the following, which tend to occur within a characteristic range:

- Methane: This is a colourless, odourless and flammable gaswith a density lighter than air, typically making up 50 – 60% of the landfill gas.
- Carbon dioxide: This is a colourless, odourless and non-inflammable gas that is denser than air, typically accounting for 30 – 40%.
- Oxygen: The flammability of methane depends on the percentage of oxygen. It is, therefore, important to control oxygen levels, where gas abstraction is undertaken.
- Nitrogen: This is essentially inert and will have little effect, except to modify the explosive range of methane.

It is difficult to convert the amount of gas measured to the maximum landfill gas production value because gas is withdrawn from a small part of the landfill only, referred to as zone of influence during measurement. In other words, it is very difficult to determine this zone and relate it to the whole landfill area.

Hazards

Landfill gas consists of a mixture of flammable, asphyxiating and noxious gases and may be hazardous to health and safety, and hence the need for precautions. Some of the major hazards are listed below:

- Explosion and fire: Methane is flammable in air within the range of 5 – 15% by volume, while hydrogen is flammable within the range of 4.1 – 7.5% (in the presence of oxygen) and potentially explosive. Fire, occurring within the waste, can be difficult to extinguish and can lead to unpredictable and uncontrolled subsidence as well as production of smoke and toxic fumes.
- Trace components: These comprise mostly alkanes and alkenes, and their oxidation products such as aldehydes, alcohols and esters . Many of them are recognised as toxicants, when present in air at concentrations above occupational exposure standards.
- Global warming: Known also as greenhouse effect, it is the warming of the earth's atmosphere by the accumulation of gases (methane, carbon dioxide and chlorofluorocarbons) that absorbs reflected solar radiation.

Migration

During landfill development, most of the gasproduced isvented to the atmosphere, provided the permeable intermediate cover has been used. While biological and chemical processes affect gas composition through methane oxidation, which converts methane to carbon dioxide, physical factors affect gas migration. The physical factors that affect gas migration include:

- Environmental conditions: These affect the rate of degradation and gas pressure build up.
- Geophysical conditions: These affect migration pathways. In the presence of fractured geological strata or a mineshaft, the gas may travel large distances, unless restricted by the water table.
- Climatic conditions: Falling atmospheric pressure, rainfall and water infiltration rate affect landfill gas migration.

The proportion of void space in the ground, rather than permeability, determines the variability of gas emission. If the escape of landfill gas is controlled and proper extraction system is designed, this gas can be utilised as a source of energy. If landfill gas is not utilised, it should be burnt by means off laring. However, landfill gas utilisation can save on the use of fossil fuels since its heating value is approximately 6 kWh/m3 and can be utilised in internal combustion engines for production of electricity and heat.

It is important that landfill gas is extracted during the operation phase. It is extracted out of the landfill by means of gas wells, which are normally drilled by auger and are driven into the landfill at a spacing of 40 – 70 m. In addition, horizontal systems can be installed during operation of the landfill. The gas wells consist mainly of perforated plastic pipes surrounded by coarse gravel and are connected with the gastransportation pipe with flexible tubing. The vacuum necessary for gas extraction and transportation is created by means of a blower. The most important factors influencing planning and construction of landfill gas extraction systems are settling of waste, water tables in landfills and gas quality.

Control

To control gas emission, it is necessary to control the following:

- waste inputs (i.e., restrict the amount of organic waste).
- processes within the waste (i.e., minimise moisture content to limit gas production).
- migration process (i.e., provide physical barriers or vents to remove the gas from the site and reduce gas pressure). Note that since gas migration cannot be easily prevented, removal is often the preferred option. This is done by using vents (extraction wells) within the waste or stone filled vents, which are often placed around the periphery of the landfill site. Some of the gas collection systems include impermeable cap, granular material, collection pipes and treatment systems.

Leachate Formation

Leachate can pollute both groundwater and surface water supplies. The degree of pollution will depend on local geology and hydrogeology, nature of waste and the proximity of susceptible receptors. Once groundwater is contaminated, it is very costly to clean it up. Landfills, therefore, undergo siting, design and construction procedures that control leachate migration.

Composition and Properties

Leachate comprises soluble components of waste and its degradation products enter water, as it percolates through the landfill. The amount of leachate generated depends on:

- water availability;
- landfill surface condition;
- refuse state;
- condition of surrounding strata.

The major factor, i.e., water availability, is affectedbyprecipitation,surface runoff, waste decomposition and liquid waste disposal. The water balance equation for landfill requires negative or zero ("Lo") so that no excess leachate is produced.This is calculated using the following formula:

$$Lo = I - E - aW$$

i.e. $I - E < aW$

where, Lo = free leachate retained at site (equivalent to leachate production minus leachate leaving the site); I = total liquid input;

E = evapotranspiration losses; a = absorption capacity of waste;

W = weight of waste disposed.

Common toxic components in leachate are ammonia and heavy metals, which can be hazardous even at low levels, if they accumulate in the food chain. The presence of ammoniacal nitrogen means that leachate often has to be treated off-site before being discharged to a sewer, since there is no natural bio-chemical path for its removal (Ali, et al., 1995). Leachate composition varies with time and location. Table shows a typical leachate properties and composition at various stages of waste decomposition:

Table:

Properties and Composition of Leachate at Various Stages of Decomposition (mg/l except pH)

Components	Fresh wastes	Aged wastes	Wastes with high moisture
pH	6.2	7.5	8.0
COD	23800	1160	1500
BOD	11900	260	500
TOC	8000	465	450
Volatile acid (as C)	5688	5	12
NH3-N	790	370	1000
NO3-N	3	1	1.0
Ortho-P	0.73	1.4	1.0
Cl	1315	2080	1390
Na	9601	300	1900
Mg	252	185	186
K	780	590	570
Ca	1820	250	158
Mn	27	2.1	0.05
Fe	540	23	2.0
Cu	0.12	0.03	-
Zn	21.5	0.4	0.5
Pb	0.40	0.14	-

Leachate Migration

It is generally difficult to predict the movement of escaped leachate accurately. The main controlling factors are the surrounding geology and hydrogeology. Escape to surface water may be relatively easy to control, but if it escapes to groundwater sources, it can be very difficult both to control and clean up. The degree of groundwater contamination is affected by physical, chemical and biological actions. The relative importance of each process may change, however, if the leachate moves from the landfill to the sub-surface region.

Control

The best way to control leachate is through prevention, which should be integral to the site design. In most cases, it is necessary to control liquid access, collection and treatment, all of which can be done using the following landfill liners:

- Natural liners: These refer to compacted clay or shale, bitumen or soil sealants, etc., and are generally less permeable, resistant to chemical attack and have good sorption properties. They generally do not act as true containment barriers, because sometimes leachate migrates through them.
- Synthetic (geo-membrane) liners: These are typically made up of high or medium density polyethylene and are generally less permeable, easy to install, relatively strong and have good deformation characteristics. They sometimes expand or shrink according to temperature and age.

Note that natural and geo-membrane liners are often combined to enhance the overall efficiency of the containment system. Some of the leachate collection systems include impermeable liner, granular material, collection piping, leachate storage tank; leachate is trucked to a wastewater treatment facility.

Treatment

Concentrations of various substances occurring in leachate are too high to be discharged to surface water or into a sewer system. These concentrations, therefore, have to be reduced by removal, treatment or both. The various treatments of leachate include:

- Leachate recirculation: It is one of the simplest forms of treatment. Recirculation of leachate reduces the hazardous nature of leachate and helps wet the waste, increasing its potential for biological degradation.
- Biological treatment: This removes BOD, ammonia and suspended solids. Leachate from land filled waste can be readily degraded by biological means, due to high content of volatile fatty acids (VFAs). The common methods are aerated lagoons (i.e., special devices which enhance the aerobic process es of degradation of organic substances over the entire depth of the tank) and activated sludge process, which differs from aerated lagoons in that discharged sludge is recirculated and is often used for BOD and ammonia removal. While under conditions of low COD, rotating biological contactors (i.e., biomassis brought into contact with circular blades fixed to a common axle which is rotated) are

very effective in removing ammonia. In an anaerobic treatment system, complex organic molecules are fermented in filter. The common types are anaerobic filters, anaerobic lagoon and digesters.

- Physicoche mical treatment: After biological degradation, effluents still contain significant concentrations of different substances. Physicochemical treatment processes could be installed to improve the leachate effluent quality. Some of these processes are flocculation-precipitation.

(Note that addition of chemicals to the water attracts the metal by floc formation). Separation of the floc from water takes place by sedimentation, adsorption and reverse osmosis.

Landfill Operation Issues

Once a potential site has been identified/selected, an assessment of design aspects, including costs for civil works, begins. Important issues to be looked into in this regard are land requirements, types of wastes that are handled, evaluation of seepage potential, design of drainage and seepage control facilities, development of a general operation plan, design of solid waste filling plan and determination of equipment requirements.

With this in view, we will discuss some important factors required for successful implementation and operation of a sanitary landfill in Subsection.

Design and Construction

The design and construction process involves site infrastructure, i.e., the position of the buildings, roads and facilities that are necessary to the efficient running of the site and site engineering, i.e., the basic engineering works needed to shape the site for the reception of wastes and to meet the technical requirements of the working plan (Phelps, 1995). At the outset, however, the potential operator and the licensing authority should agree upon a working plan for the landfill. The disposal license includes the design, earthworks and procedures in the working plan.

What are the processes involved in design and construction? We will study these below:

(i) Site infrastructure: The size, type and number of buildings required at a landfill depend on factors such as the level of waste input, the expected life of the site and environmental factors. Depending on the size and complexity of the landfill, buildings range from single portable cabins to big complexes. However, certain aspects such as the following are common:

- o need to comply with planning, building, fire, healthandsafety regulations and controls;
- o security and resistance to vandalism;
- o durability of service and the possible need to relocate accommodation during the lifetime of the site operations;
- o ease of cleaning and maintenance;

- o availability of services such as electricity, water, drainage and telecommunication.

Paying some attention to the appearance of the site entrance is necessary, as it influences the perception of the public about the landfill site. All landfill sites need to control and keep records of vehicles entering and leaving the site, and have a weighbridge to record waste input data, which can be analysed by a site control office. Note that at small sites, the site control office can be accommodated at the site itself.

(ii) Earthworks: Various features of landfill operations may require substantial earthworks, and therefore, the working plan must include earthworks to be carried out before wastes can be deposited. Details about earthworks gain significance, if artificial liners are to be installed, which involves grading the base and sides of the site (including construction of 25 slopes to drain leachate to the collection areas) and the formation of embankments.Material may also have to be placed in stockpiles for later use at the site. The cell method of operation requires the construction of cell walls. At some sites, it may be necessary to construct earth banks around the site perimeter to screen the landfill operations from the public. Trees or shrubs may then be planted on the banks to enhance the screening effect. The construction of roads leading to disposal sites also involves earthworks.

(iii) Lining landfill sites: Where the use of a liner is envisaged, the suitability of a site for lining should be evaluated at the site investigation stage. However, they should not be installed, until the site has been properly prepared.The areato be lined should be free of objects likely to cause physical damage to the liner, such as vegetation and hard rocks. If synthetic liner materials are used, a binding layer of suitable fine-grained material should be laid to support the liner. However, if the supporting layer consists of low permeable material (e.g., clay), the synthetic liner must be placed on top of this layer. A layer of similar fine- grained material with the thickness of 25 – 30 cm should also be laid above the liner to protect it from subsequent mechanical and environmental damage. During the early phase of operation, particular care should be taken to ensure that the traffic does not damage the liner. Monitoring the quality of groundwater close to the site is necessary to get the feedback on the performance of a liner.

(iv) Leachate and landfill gas management: The basic elements of the leachate collection system (i.e., drain pipes, drainage layers, collection pipes, sumps, etc.) must be installed immediately above the liner, before any waste is deposited. Particular care must also be taken to prevent the drain and collection pipes from settling. During landfill operations, waste cells are covered with soil to avoid additional contact between waste and the environment. The soil layers have to be sufficiently permeable to allow downward leachate transport. Landfill gas is not extracted before completion, which includes construction of final cover, of the waste body. Extraction wells (diameter 0.3 to 1.0 m) may be constructed during or after operation.

(v) Landfill capping: Capping is required to control and minimise leachate generation (by minimising water ingress into the landfill) and facilitate landfill gas control or collection (by installing a low permeability cap over the whole site). A cap may consist of natural (e.g., clay) or synthetic (e.g., poly-ethylene) material with thickness of at least 1 m. An uneven settlement of the waste may be a major cause of cap failure. Designs for capping should, therefore, include consideration of leachate and landfill gas collection wells or vents. For the cap toremain effective, it must be protected from agricultural machinery, drying and cracking, plant root penetration, burrowing animals and erosion.

Operation

To secure public acceptability, landfill operations require careful planning and determination of the extent of environmental effects. The basic factor influencing the planning of site operations is the nature and quantity of incoming wastes. The various aspects of this include the following:

(i) Methods of filling: The following variations in land filling techniques are available (Burner and Kelly, 1972):

- Trench method: This involves the excavation of a trench into which waste is deposited, and the excavated material is then used as cover.
- Area method: Wastes may be deposited in layers and so form terraces over the available area. However, with this type of operation, excessive leachate generation may occur, which may render the control difficult.
- Cell method: This method involves the deposition of wastes within pre- constructed bounded area. It is now the preferred method in the industrialised world, since it encourages the concept of progressive filling and restoration. Operating a cellular method of filling enables wastes to be deposited in a tidy manner, as the cells serve both to conceal the tipping operation and trap much of the litter that has been generated.
- Canyon/depression: This method refers to the placing of suitable wastes against lined canyon or ravine slide slopes. (Slope stability and leachate gas emission control are critical issues for this type of waste placement.)

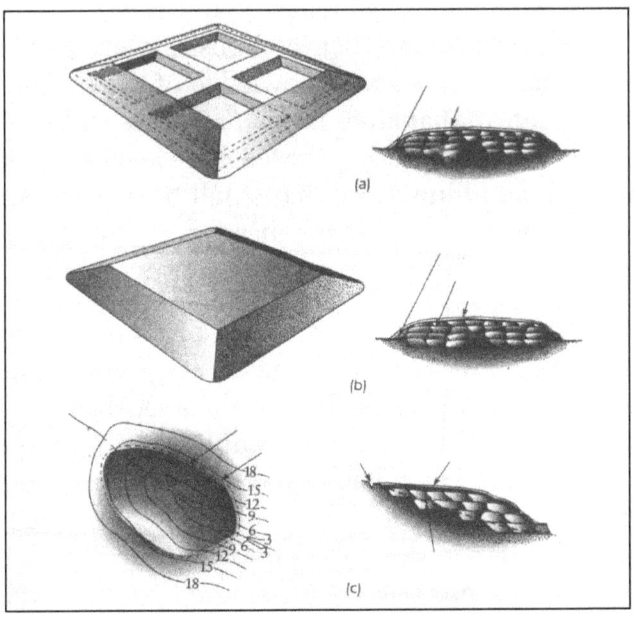

Commonly Used Land Filling Methods.
(a) trench (b) areaand (c) canyon/depression methods.

(ii) Refuse placement: The working space should be sufficiently extensive to permit vehicles to manoeuvre and unload quickly and safely without impeding refuse spreading, and allow easy operation of the site equipment. Depositing waste in thin layers and using a compactor enables a high waste density to be achieved. Each progressive layer should not be more than 30 cm thick. The number of passes by a machine over the waste determines the level of compaction.

(iii) Covering of waste: At the end of each working day, all exposed surfaces, including the flanks and working space, should be covered witha suitable inert material to a depth of at least 15 cm. This daily cover is considered essential, as it minimises wind blown litter and helps reduce odours. Cover material may be obtained from on-site excavations or inert waste materials coming to the site. Pulverised fuel ash or sewage sludge can also be used for this purpose.

(iv) Site equipment and workforce orientation: The equipment most commonly used on landfill sites includes steel wheeled compactors, tracked dozers, loaders, earthmovers and hydraulic excavators. Scrapers are used for excavating and moving cover materials. In addition to appropriate equipment, proper training must be ensured for the workforce. They should be competent, and adequately supervised; training should include site safetyandfirstaid. Since a landfill site may pose dangers to both site operators and users, it is necessary to lay down emergency plans and test them from time to time (Phelps, 1995).

Monitoring

Landfill represents a complex process of transforming polluting wastes into environmentally acceptable deposits. Because of the complexity of these processes and their potential environmental effects, it is imperative to monitor and confirm that the landfill works, as expected. A monitoring scheme, for example, is required for collecting detailed information on the development of leachate and landfill gas within and beyond a landfill. The scheme should be site specific, drawn at the site investigation stage and implemented. Monitoring is generally done for the following:

(i) Leachate/gas: Monitoring of leachate/gas plays a vital role in the management of landfills. Data on the volume of leachate/gas and their composition are essential for proper control of leachate/gas generation and its treatment. Knowledge of the chemical composition of leachate/gas is also required to confirm that attenuation processes within the landfill are proceeding as expected. Various systems for monitoring the leachate level are in use, and are mostly based on pipes installed prior to land filling. Note that small bore perforated plastic pipes are relatively cheaper and easier to install, but have the disadvantage of getting damaged faster during infilling. Placing pipes within a column or tyres may, however, offer some protection.

(ii) Groundwater: A continued groundwater-monitoring programme for confirming the integrity of the liner system is essential. At an early stage of site preparation, therefore, a number of monitoring boreholes need to be provided around the site. However, the location, design and number of boreholes depend on the size of the landfill, proximity to an aquifer, geology of the site and types of wastes deposited. Installation of a double liner system can make the monitoring exercise more accurate and easier to perform. Water should be regularly flushed through the secondary leachate collection system. In case this water is polluted, the primary leachate barrier will be damaged, and if repair is not considered possible, the leachate collected must be transported to the leachate treatment facility.

Key Issues in Waste Disposal

Let us first get one thing very clear: there is no option but to dispose of wastes. Disposal is the final element in the SWM system. It is the ultimate fate of all solid wastes, be they residential wastes collected and transported directly to a landfill site, semisolid waste (sludge) from municipal and industrial treatment plants, incinerator residue, compost or other substances from various solid waste processing plants that are of no further use to society. It is, therefore, imperative to have a proper plan in place for safe disposal of solid wastes, which involves appropriate handling of residual matter after solid wastes have been processed and the recovery of conversion products/energy has been achieved. It follows that an efficient SWM system must provide an environmentally sound disposal option for waste that cannot be reduced, recycled, composted, combusted, or processed further (Ali, et al 1999).

However, in these days, indiscriminate disposal of wastesin many regionsis very common, giving rise to such problems as:

- health hazards (e.g., residents in the vicinity of wastes inhale dust and smoke when the wastes are burnt; workers and rag pickers come into direct contact with wastes, etc.);
- pollution due to smoke;
- pollution from waste leachate and gas;
- blockage of open drains and sewers.

clearly, safe disposal of solid wastes is important for safeguarding both public health and the environment.

Issues to be Overcome

To achieve effective waste disposal, we must overcome the following the constraints:

(i) Municipal capacities: With the increasing volume of waste generation, collection of wastes gets more attention than disposal. Furthermore, in India, only a few municipalities seem to have the required experience or capacity for controlled disposal. Some municipalities may have identified disposal sites but still only few may actively manage them. In some places, contracting out waste disposal is seen as a solution. But, municipalities are not equipped to deal with the problems associated with it, such as issues of privatisation and monitoring of the contract.

(ii) Political commitment: SWM is more than a technical issue, as any successful programme needs effective political and governmental support. This is rarely a priority of government authorities, unless there is a strong and active public interest as well as international interventions.

(iii) Finance and cost recovery: Development of a sanitary landfill site represents a major investment and it generally receives less priority over other resource demands. And, even when establishment costs are secured for a disposal site, recurrent costs to maintain it always pose problems.

(iv) Technical guidelines: Standards established for waste disposal in one country need not necessarily be appropriate for another, due to reasons such as climatic conditions, resources availability, institutional infrastructure, socio-cultural values, etc. In the absence of adequate data and/or the means of collecting/acquiring it, officials often struggle to plan a safe and economically viable disposal option.

(v) Institutional role and responsibility: A disposal site may be located outside the boundary of a town and may serve more than one town. This necessitates the co-ordination of all authorities concerned, and the roles and responsibilities of different departments need to be clearly defined and accepted by all concerned.

(vi) Location: The accessibility of a disposal site, especially its distance from town, is an important factor in site selection, especially when staff and public do not have a strong incentive to use it, when compared to indiscriminate dumping. Site selection is perhaps the most difficult stage in the development of suitable disposal option.

Chapter 5

Waste Recycling

The process of conversion of waste materials into new and useful objects is known as recycling. This chapter explains some aspects of recycling such as plastic recycling, computer recycling, ship breaking, ferrous metal recycling and precycling in an easy-to-understand manner. The topics elaborated in this chapter will help in gaining a better perspective about recycling.

Recycling

The three chasing arrows of the international recycling logo. It is sometimes accompanied by the text "reduce, reuse and recycle".

Recycling is the process of converting waste materials into reusable objects to prevent waste of potentially useful materials, reduce the consumption of fresh raw materials, energy usage, air pollution (from incineration) and water pollution (from landfilling) by decreasing the need for "conventional" waste disposal and lowering greenhouse gas emissions compared to plastic production. Recycling is a key component of modern waste reduction and is the third component of the "Reduce, Reuse and Recycle" waste hierarchy.

There are some ISO standards related to recycling such as ISO 15270:2008 for plastics waste and ISO 14001:2004 for environmental management control of recycling practice.

Recyclable materials include many kinds of glass, paper and cardboard, metal, plastic, tires, textiles and electronics. The composting or other reuse of biodegradable waste—such as food or garden waste—is also considered recycling. Materials to be recycled are either brought to a collection centre or picked up from the curbside, then sorted, cleaned and reprocessed into new materials destined for manufacturing.

In the strictest sense, recycling of a material would produce a fresh supply of the same material—for example, used office paper would be converted into new office paper, or used polystyrene foam into new polystyrene. However, this is often difficult or too expensive (compared with producing the same product from raw materials or other sources), so "recycling" of many products or materials involves their *reuse* in producing different materials (for example, paperboard) instead. Another form of recycling is the salvage of certain materials from complex products, either due to their intrinsic value (such as lead from car batteries, or gold from circuit boards), or due to their hazardous nature (e.g., removal and reuse of mercury from thermometers and thermostats).

History

Origins

Recycling has been a common practice for most of human history, with recorded advocates as far back as Plato in 400 BC. During periods when resources were scarce, archaeological studies of ancient waste dumps show less household waste (such as ash, broken tools and pottery)—implying more waste was being recycled in the absence of new material.

An American poster from World War II.

In pre-industrial times, there is evidence of scrap bronze and other metals being collected in Europe and melted down for perpetual reuse. Paper recycling was first recorded in 1031, when Japanese shops sold repulped paper. In Britain dust and ash from wood and coal fires was collected by "dustmen" and downcycled as a base material used in brick making. The main driver for these types of recycling was the economic advantage of obtaining recycled feedstock instead of acquiring virgin material, as well as a lack of public waste removal in ever more densely populated areas. In 1813, Benjamin Law developed the process of turning rags into "shoddy" and "mungo" wool in

Batley, Yorkshire. This material combined recycled fibers with virgin wool. The West Yorkshire shoddy industry in towns such as Batley and Dewsbury, lasted from the early 19th century to at least 1914.

Industrialization spurred demand for affordable materials; aside from rags, ferrous scrap metals were coveted as they were cheaper to acquire than virgin ore. Railroads both purchased and sold scrap metal in the 19th century, and the growing steel and automobile industries purchased scrap in the early 20th century. Many secondary goods were collected, processed and sold by peddlers who scoured dumps and city streets for discarded machinery, pots, pans and other sources of metal. By World War I, thousands of such peddlers roamed the streets of American cities, taking advantage of market forces to recycle post-consumer materials back into industrial production.

Beverage bottles were recycled with a refundable deposit at some drink manufacturers in Great Britain and Ireland around 1800, notably Schweppes. An official recycling system with refundable deposits was established in Sweden for bottles in 1884 and aluminum beverage cans in 1982; the law led to a recycling rate for beverage containers of 84–99 percent depending on type, and a glass bottle can be refilled over 20 times on average.

British poster from World War II.

Wartime

New chemical industries created in the late 19th century both invented new materials (e.g. Bakelite (1907) and promised to transform valueless into valuable materials. Proverbially, you could not make a silk purse of a sow's ear—until the US firm Arhur D. Little published in 1921 "On the Making of Silk Purses from Sows' Ears", its research proving that when "chemistry puts on overalls and gets down to business . . .new values appear. New and better paths are opened to reach the goals desired."

Recycling was a highlight throughout World War II. During the war, financial constraints and significant material shortages due to war efforts made it necessary for countries to reuse goods and recycle materials. These resource shortages caused by the world wars, and other such world-changing occurrences, greatly encouraged recycling. The struggles of war claimed much of the material resources available, leaving little for the civilian population. It became necessary for most homes to recycle their waste, as recycling offered an extra source of materials allowing people to make the most of what was available to them. Recycling household materials meant more resources for war efforts and a better chance of victory. Massive government promotion campaigns were carried out in the home front during World War II in every country involved in the war, urging citizens to donate metals and conserve fiber, as a matter of patriotism.

Post-war

A considerable investment in recycling occurred in the 1970s, due to rising energy costs. Recycling aluminum uses only 5% of the energy required by virgin production; glass, paper and metals have less dramatic but very significant energy savings when recycled feedstock is used.

Although consumer electronics such as the television have been popular since the 1920s, recycling of them was almost unheard of until early 1991. The first electronic waste recycling scheme was implemented in Switzerland, beginning with collection of old refrigerators but gradually expanding to cover all devices. After these schemes were set up, many countries did not have the capacity to deal with the sheer quantity of e-waste they generated or its hazardous nature. They began to export the problem to developing countries without enforced environmental legislation. This is cheaper, as recycling computer monitors in the United States costs 10 times more than in China. Demand in Asia for electronic waste began to grow when scrap yards found that they could extract valuable substances such as copper, silver, iron, silicon, nickel and gold, during the recycling process. The 2000s saw a large increase in both the sale of electronic devices and their growth as a waste stream: in 2002, e-waste grew faster than any other type of waste in the EU. This caused investment in modern, automated facilities to cope with the influx of redundant appliances, especially after strict laws were implemented in 2003.

As of 2014, the European Union has about 50% of world share of the waste and recycling industries, with over 60,000 companies employing 500,000 persons, with a turnover of €24 billion. Countries have to reach recycling rates of at least 50%, while the lead countries are around 65% and the EU average is 39% as of 2013.

Legislation

Supply

For a recycling program to work, having a large, stable supply of recyclable material is crucial. Three legislative options have been used to create such a supply: mandatory recycling collection, container deposit legislation and refuse bans. Mandatory collection laws set recycling targets for cities to aim for, usually in the form that a certain percentage of a material must be diverted from the city's waste stream by a target date. The city is then responsible for working to meet this target.

Container deposit legislation involves offering a refund for the return of certain containers, typically glass, plastic and metal. When a product in such a container is purchased, a small surcharge is added to the price. This surcharge can be reclaimed by the consumer if the container is returned to a collection point. These programs have been very successful, often resulting in an 80 percent recycling rate. Despite such good results, the shift in collection costs from local government to industry and consumers has created strong opposition to the creation of such programs in some areas. A variation on this is where the manufacturer bears responsibility for the recycling of their goods. In the European Union, the WEEE Directive requires producers of consumer electronics to reimburse the recyclers' costs.

An alternative way to increase supply of recyclates is to ban the disposal of certain materials as waste, often including used oil, old batteries, tires and garden waste. One aim of this method is to create a viable economy for proper disposal of banned products. Care must be taken that enough of these recycling services exist, or such bans simply lead to increased illegal dumping.

Government-mandated Demand

Legislation has also been used to increase and maintain a demand for recycled materials. Four methods of such legislation exist: minimum recycled content mandates, utilization rates, procurement policies and recycled product labeling.

Both minimum recycled content mandates and utilization rates increase demand directly by forcing manufacturers to include recycling in their operations. Content mandates specify that a certain percentage of a new product must consist of recycled material. Utilization rates are a more flexible option: industries are permitted to meet the recycling targets at any point of their operation or even contract recycling out in exchange for tradeable credits. Opponents to both of these methods point to the large increase in reporting requirements they impose, and claim that they rob industry of necessary flexibility.

Governments have used their own purchasing power to increase recycling demand through what are called "procurement policies." These policies are either "set-asides," which reserve a certain amount of spending solely towards recycled products, or "price preference" programs which provide a larger budget when recycled items are purchased. Additional regulations can target specific cases: in the United States, for example, the Environmental Protection Agency mandates the purchase of oil, paper, tires and building insulation from recycled or re-refined sources whenever possible.

The final government regulation towards increased demand is recycled product labeling. When producers are required to label their packaging with amount of recycled material in the product (including the packaging), consumers are better able to make educated choices. Consumers with sufficient buying power can then choose more environmentally conscious options, prompt producers to increase the amount of recycled material in their products, and indirectly increase demand. Standardized recycling labeling can also have a positive effect on supply of recyclates if the labeling includes information on how and where the product can be recycled.

Recyclates

Glass recovered by crushing only one kind of beer bottle.

Recyclate is a raw material that is sent to, and processed in a waste recycling plant or materials recovery facility which will be used to form new products. The material is collected in various methods and delivered to a facility where it undergoes re-manufacturing so that it can used in the production of new materials or products. For example, plastic bottles that are collected can be re-used and made into plastic pellets, a new product.

Quality of Recyclate

The quality of recyclates is recognized as one of the principal challenges that needs to be addressed for the success of a long-term vision of a green economy and achieving zero waste. Recyclate quality is generally referring to how much of the raw material is made up of target material compared to the amount of non-target material and other non-recyclable material. Only target material is likely to be recycled, so a higher amount of non-target and non-recyclable material will reduce the quantity of recycling product. A high proportion of non-target and non-recyclable material can make it more difficult for re-processors to achieve "high-quality" recycling. If the recyclate is of poor quality, it is more likely to end up being down-cycled or, in more extreme cases, sent to other recovery options or landfilled. For example, to facilitate the re-manufacturing of clear glass products there are tight restrictions for colored glass going into the re-melt process.

The quality of recyclate not only supports high quality recycling, but it can also deliver significant environmental benefits by reducing, reusing and keeping products out of landfills. High quality recycling can help support growth in the economy by maximizing the economic value of the waste material collected. Higher income levels from the sale of quality recyclates can return value which can be significant to local governments, households and businesses. Pursuing high quality recycling can also provide consumer and business confidence in the waste and resource management sector and may encourage investment in that sector.

There are many actions along the recycling supply chain that can influence and affect the material quality of recyclate. It begins with the waste producers who place non-target and non-recyclable wastes in recycling collection. This can affect the quality of final recyclate streams or require further efforts to discard those materials at later stages in the recycling process. The different collection systems can result in different levels of contamination. Depending on which materials are collected together, extra effort is required to sort this material back into separate streams and can significantly reduce the quality of the final product. Transportation and the compaction of materials can make it more difficult to separate material back into separate waste streams. Sorting facilities are not one hundred per cent effective in separating materials, despite improvements in technology and quality recyclate which can see a loss in recyclate quality. The storage of materials outside where the product can become wet can cause problems for re-processors. Reprocessing facilities may require further sorting steps to further reduce the amount of non-target and non-recyclable material. Each action along the recycling path plays a part in the quality of recyclate.

Quality Recyclate Action Plan (Scotland)

The Recyclate Quality Action Plan of Scotland sets out a number of proposed actions that the Scottish Government would like to take forward in order to drive up the quality of the materials being collected for recycling and sorted at materials recovery facilities before being exported or sold on to the reprocessing market.

The plan's objectives are to:

- Drive up the quality of recyclate.
- Deliver greater transparency about the quality of recyclate.
- Provide help to those contracting with materials recycling facilities to identify what is required of them.
- Ensure compliance with the Waste (Scotland) regulations 2012.
- Stimulate a household market for quality recyclate.
- Address and reduce issues surrounding the Waste Shipment Regulations.

The plan focuses on three key areas, with fourteen actions which were identified to increase the quality of materials collected, sorted and presented to the processing market in Scotland.

The three areas of focus are:

- Collection systems and input contamination.
- Sorting facilities – material sampling and transparency.
- Material quality benchmarking and standards.

Recycling Consumer Waste

Collection

A number of different systems have been implemented to collect recyclates from the general waste stream. These systems lie along the spectrum of trade-off between public convenience and government ease and expense. The three main categories of collection are "drop-off centers," "buy-back centers" and "curbside collection."

A three-sided bin at a railway station in Germany, intended to separate paper *(left)* and plastic wrappings *(right)* from other waste *(back)*.

Curbside Collection

Curbside collection encompasses many subtly different systems, which differ mostly on where in the process the recyclates are sorted and cleaned. The main categories are mixed waste collection, commingled recyclables and source separation. A waste collection vehicle generally picks up the waste.

At one end of the spectrum is mixed waste collection, in which all recyclates are collected mixed in with the rest of the waste, and the desired material is then sorted out and cleaned at a central sorting facility. This results in a large amount of recyclable waste, paper especially, being too soiled to reprocess, but has advantages as well: the city need not pay for a separate collection of recyclates and no public education is needed. Any changes to which materials are recyclable is easy to accommodate as all sorting happens in a central location.

A recycling truck collecting the contents of a recycling bin in Canberra, Australia.

In a commingled or single-stream system, all recyclables for collection are mixed but kept separate from other waste. This greatly reduces the need for post-collection cleaning but does require public education on what materials are recyclable.

Source separation is the other extreme, where each material is cleaned and sorted prior to collection. This method requires the least post-collection sorting and produces the purest recyclates, but incurs additional operating costs for collection of each separate material. An extensive public education program is also required, which must be successful if recyclate contamination is to be avoided.

Source separation used to be the preferred method due to the high sorting costs incurred by commingled (mixed waste) collection. Advances in sorting technology however, have lowered this overhead substantially—many areas which had developed source separation programs have since switched to co-mingled collec-tion.

Buy-back Centers

Buy-back centers differ in that the cleaned recyclates are purchased, thus providing a clear incentive for use and creating a stable supply. The post-processed material can then be sold on, hopefully creating a profit. Unfortunately, government subsidies are necessary to make buy-back centres a viable enterprise, as according to the U.S. National Waste & Recycling Association, it costs on average US$50 to process a ton of material, which can only be resold for US$30.

Drop-off Centers

Drop-off centers require the waste producer to carry the recyclates to a central location, either an installed or mobile collection station or the reprocessing plant itself. They are the easiest type of collection to establish, but suffer from low and unpredictable throughput.

Distributed Recycling

For some waste materials such as plastic, recent technical devices called recyclebots enable a form of distributed recycling. Preliminary life-cycle analysis (LCA) indicates that such distributed recycling of HDPE to make filament of 3-D printers in rural regions is energetically favorable to either using virgin resin or conventional recycling processes because of reductions in transportation energy.

Sorting

Once commingled recyclates are collected and delivered to a central collection facility, the different types of materials must be sorted. This is done in a series of stages, many of which involve automated processes such that a truckload of material can be fully sorted in less than an hour. Some plants can now sort the materials automatically, known as single-stream recycling. In plants, a variety of materials are sorted such as paper, different types of plastics, glass, metals, food scraps and most types of batteries. A 30 percent increase in recycling rates has been seen in the areas where these plants exist.

Recycling sorting facility and processes.

Initially, the commingled recyclates are removed from the collection vehicle and placed on a conveyor belt spread out in a single layer. Large pieces of corrugated fiberboard and plastic bags are removed by hand at this stage, as they can cause later machinery to jam.

Early sorting of recyclable materials: glass and plastic bottles in Poland.

Next, automated machinery such as disk screens and air classifiers separate the recyclates by weight, splitting lighter paper and plastic from heavier glass and metal. Cardboard is removed from the mixed paper and the most common types of plastic, PET (#1) and HDPE (#2), are collected. This separation is usually done by hand but has become automated in some sorting centers: a spectroscopic scanner is used to differentiate between different types of paper and plastic based on the absorbed wavelengths, and subsequently divert each material into the proper collection channel.

Strong magnets are used to separate out ferrous metals, such as iron, steel and tin cans. Non-ferrous metals are ejected by magnetic eddy currents in which a rotating magnetic field induces an electric current around the aluminum cans, which in turn creates a magnetic eddy current inside the cans. This magnetic eddy current is repulsed by a large magnetic field, and the cans are ejected from the rest of the recyclate stream.

Finally, glass is sorted according to its color: brown, amber, green or clear. It may either be sorted by hand, or via an automated machine that uses colored filters to detect different colors. Glass fragments smaller than 10 millimetres (0.39 in) across cannot be sorted automatically, and are mixed together as "glass fines."

A recycling point in New Byth, Scotland, with separate containers for paper, plastics and differently colored glass.

This process of recycling as well as reusing the recycled material has proven advantageous because it reduces amount of waste sent to landfills, conserves natural resources, saves energy, reduces greenhouse gas emissions and helps create new jobs. Recycled materials can also be converted into new products that can be consumed again, such as paper, plastic and glass.

The City and County of San Francisco's Department of the Environment is attempting to achieve a citywide goal of Zero Waste by 2020. San Francisco's refuse hauler, Recology, operates an effective recyclables sorting facility in San Francisco, which helped San Francisco reach a record-breaking diversion rate of 80%.

Rinsing

Food packaging should no longer contain any organic matter (organic matter, if any, needs to be placed in a biodegradable waste bin or be buried in a garden). Since no trace of biodegradable material is best kept in the packaging before placing it in a trash bag, some packaging also needs to be rinsed.

Recycling Industrial Waste

Although many government programs are concentrated on recycling at home, a 64% of waste in the United Kingdom is generated by industry. The focus of many recycling programs done by industry is the cost–effectiveness of recycling. The ubiquitous nature of cardboard packaging makes cardboard a commonly recycled waste product by companies that deal heavily in packaged goods, like retail stores, warehouses and distributors of goods. Other industries deal in niche or specialized products, depending on the nature of the waste materials that are present.

The glass, lumber, wood pulp and paper manufacturers all deal directly in commonly recycled materials; however, old rubber tires may be collected and recycled by independent tire dealers for a profit.

Mounds of shredded rubber tires are ready for processing.

Levels of metals recycling are generally low. In 2010, the International Resource Panel, hosted by the United Nations Environment Programme (UNEP) published reports on metal stocks that exist within society and their recycling rates. The Panel reported that the increase in the use of metals during the 20th and into the 21st century has led to a substantial shift in metal stocks from below ground to use in applications within society above ground. For example, the in-use stock of copper in the USA grew from 73 to 238 kg per capita between 1932 and 1999.

The report authors observed that, as metals are inherently recyclable, the metal stocks in society can serve as huge mines above ground (the term "**urban mining**" has been coined with this idea in mind). However, they found that the recycling rates of many metals are very low. The report warned that the recycling rates of some rare metals used in applications such as mobile phones, battery packs for hybrid cars and fuel cells, are so low that unless future end-of-life recycling rates are dramatically stepped up these critical metals will become unavailable for use in modern technology.

Aerial photo of a ship recycling facility in Chittagong, Bangladesh.

The military recycles some metals. The U.S. Navy's Ship Disposal Program uses ship breaking to reclaim the steel of old vessels. Ships may also be sunk to create an artificial reef. Uranium is a very dense metal that has qualities superior to lead and titanium for many military and industrial uses. The uranium left over from processing it into nuclear weapons and fuel for nuclear reactors is called depleted uranium, and it is used by all branches of the U.S. military use for armour-piercing shells and shielding.

The construction industry may recycle concrete and old road surface pavement, selling their waste materials for profit.

Some industries, like the renewable energy industry and solar photovoltaic technology in particular, are being proactive in setting up recycling policies even before there is considerable volume to their waste streams, anticipating future demand during their rapid growth.

Recycling of plastics is more difficult, as most programs are not able to reach the necessary level of quality. Recycling of PVC often results in downcycling of the material, which means only products of lower quality standard can be made with the recycled material. A new approach which allows an equal level of quality is the Vinyloop process. It was used after the London Olympics 2012 to fulfill the PVC Policy.

E-waste Recycling

E-waste is a growing problem, accounting for 20-50 million metric tons of global waste per year according to the EPA. It is also the fastest growing waste stream in the EU. Many recyclers do not recycle e-waste responsibly. After the cargo barge Khian Sea dumped 14,000 metric tons of toxic ash in Haiti, the Basel Convention was formed to stem the flow of hazardous substances into poorer countries. They created the e-Stewards certification to ensure that recyclers are held to the highest standards for environmental responsibility and to help consumers identify responsible recyclers. This works alongside other prominent legislation, such as the Waste Electrical and Electronic Equipment Directive of the EU the United States National Computer Recycling Act, to prevent poisonous chemicals from entering waterways and the atmosphere.

Microprocessors retrieved from waste stream.

In the recycling process, television sets, monitors, cell phones and computers are typically tested for reuse and repaired. If broken, they may be disassembled for parts still having high value if labor is cheap enough. Other e-waste is shredded to pieces roughly 10 centimetres (3.9 in) in size, and manually checked to separate out toxic batteries and capacitors which contain poisonous metals. The remaining pieces are further shredded to 10 millimetres (0.39 in) particles and passed under a magnet to remove ferrous metals. An eddy current ejects non-ferrous metals, which are sorted by density either by a centrifuge or vibrating plates. Precious metals can be dissolved in acid, sorted, and smelted into ingots. The remaining glass and plastic fractions are separated by density and sold to re-processors. Television sets and monitors must be manually disassembled to remove lead from CRTs or the mercury backlight from LCDs.

Plastic Recycling

Plastic recycling is the process of recovering scrap or waste plastic and reprocessing the material into useful products, sometimes completely different in form from their original state. For instance, this could mean melting down soft drink bottles and then casting them as plastic chairs and tables.

A container for recycling used plastic spoons into material for 3D printing.

Physical Recycling

Some plastics are remelted to form new plastic objects; for example, PET water bottles can be converted into polyester destined for clothing. A disadvantage of this type of recycling is that the molecular weight of the polymer can change further and the levels of unwanted substances in the plastic can increase with each remelt.

Chemical Recycling

For some polymers, it is possible to convert them back into monomers, for example PET can be treated with an alcohol and a catalyst to form a dialkyl terephthalate. The terephthalate diester can be used with ethylene glycol to form a new polyester polymer, thus making it possible to use the pure polymer again.

Waste Plastic Pyrolysis to Fuel Oil

Another process involves conversion of assorted polymers into petroleum by a much less precise thermal depolymerization process. Such a process would be able to accept almost any polymer or mix of polymers, including thermoset materials such as vulcanized rubber tires and the biopolymers in feathers and other agricultural waste. Like natural petroleum, the chemicals produced can be used as fuels or as feedstock. A RESEM Technology plant of this type in Carthage, Missouri, USA, uses turkey waste as input material. Gasification is a similar process, but is not technically recycling since polymers are not likely to become the result. Plastic Pyrolysis can convert petroleum based waste streams such as plastics into quality fuels, carbons. Given below is the list of suitable plastic raw materials for pyrolysis:

- Mixed plastic (HDPE, LDPE, PE, PP, Nylon, Teflon, PS, ABS, FRP, etc.)
- Mixed waste plastic from waste paper mill
- Multi-layered plastic

Recycling Codes

In order to meet recyclers' needs while providing manufacturers a consistent, uniform system, a coding system was developed. The recycling code for plastics was introduced in 1988 by the plastics industry through the Society of the Plastics Industry. Because municipal recycling programs traditionally have targeted packaging—primarily bottles and containers—the resin coding system offered a means of identifying the resin content of bottles and containers commonly found in the residential waste stream.

Recycling codes on products.

Plastic products are printed with numbers 1–7 depending on the type of resin. Type 1 (polyethylene terephthalate) is commonly found in soft drink and water bottles. Type 2 (high-density polyethylene) is found in most hard plastics such as milk jugs, laundry detergent bottles and some dishware. Type 3 (polyvinyl chloride) includes items such as shampoo bottles, shower curtains, hula hoops, credit cards, wire jacketing, medical equipment, siding and piping. Type 4 (low-density polyethylene) is found in shopping bags, squeezable bottles, tote bags, clothing, furniture and carpet. Type 5 is polypropylene and makes up syrup bottles, straws, Tupperware and some automotive parts. Type 6 is polystyrene and makes up meat trays, egg cartons, clamshell containers and compact disc cases. Type 7 includes all other plastics such as bulletproof materials, 3- and 5-gallon water bottles and sunglasses. Having a recycling code or the chasing arrows logo on a material is not an automatic indicator that a material is recyclable but rather an explanation of what the material is. Types 1 and 2 are the most commonly recycled.

Economic Impact

Critics dispute the net economic and environmental benefits of recycling over its costs, and suggest that proponents of recycling often make matters worse and suffer from confirmation bias. Specifically, critics argue that the costs and energy used in collection and transportation detract from (and outweigh) the costs and energy saved in the production process; also that the jobs produced by the recycling industry can be a poor trade for the jobs lost in logging, mining, and other industries associated with production; and that materials such as paper pulp can only be recycled a few times before material degradation prevents further recycling.

The National Waste and Recycling Association (NWRA), reported in May 2015, that recycling and waste made a $6.7 billion economic impact in Ohio, U.S., and employed 14,000 people.

Cost–benefit Analysis

Environmental effects of recycling		
Material	**Energy savings**	**Air pollution savings**
Aluminium	95%	95%
Cardboard	24%	—
Glass	5–30%	20%
Paper	40%	73%
Plastics	70%	—
Steel	60%	—

There is some debate over whether recycling is economically efficient. It is said that dumping 10,000 tons of waste in a landfill creates six jobs, while recycling 10,000 tons of waste can create over 36 jobs. However, the cost effectiveness of creating the additional jobs remains unproven. According to the U.S. Recycling Economic Informational Study, there are over 50,000 recycling establishments that have created over a million jobs in the US. Two years after New York City declared that implementing recycling programs would be "a drain on the city," New York City leaders realized that an efficient recycling system could save the city over $20 million. Municipalities often see fiscal benefits from implementing recycling programs, largely due to the reduced landfill costs. A study conducted by the Technical University of Denmark according to the Economist found that in 83 percent of cases, recycling is the most efficient method to dispose of household waste. However, a 2004 assessment by the Danish Environmental Assessment Institute concluded that incineration was the most effective method for disposing of drink containers, even aluminium ones.

Fiscal efficiency is separate from economic efficiency. Economic analysis of recycling does not include what economists call externalities, which are unpriced costs and benefits that accrue to individuals outside of private transactions. Examples include: decreased air pollution and greenhouse gases from incineration, reduced hazardous waste leaching from landfills, reduced energy consumption, and reduced waste and resource consumption, which leads to a reduction in environmentally damaging mining and timber activity. About 4,000 minerals are known, of these only a few hundred minerals in the world are relatively common. Known reserves of phosphorus will be exhausted within the next 100 years at current rates of usage. Without mechanisms such as taxes or subsidies to internalize externalities, businesses will ignore them despite the costs imposed on society. To make such nonfiscal benefits economically relevant, advocates have pushed for legislative action to increase the demand for recycled materials. The United States Environmental Protection Agency (EPA) has concluded in favor of recycling, saying that recycling efforts reduced the country's carbon emissions by a net 49 million metric tonnes in 2005. In the United Kingdom, the Waste and Resources Action Programme stated that Great Britain's recycling efforts reduce CO_2 emissions by 10–15 million tonnes a year. Recycling is more efficient in densely populated areas, as there are economies of scale involved.

Certain requirements must be met for recycling to be economically feasible and environmentally effective. These include an adequate source of recyclates, a system to extract those recyclates from the waste stream, a nearby factory capable of reprocessing the recyclates, and a potential demand for the recycled products. These last two requirements are often overlooked—without both an industrial market for production using the collected materials and a consumer market for the manufactured goods, recycling is incomplete and in fact only "collection".

Free-market economist Julian Simon remarked "There are three ways society can organize waste disposal: (a) commanding,(b) guiding by tax and subsidy, and (c) leaving it to the individual and the market". These principles appear to divide economic thinkers today.

Frank Ackerman favours a high level of government intervention to provide recycling services. He believes that recycling's benefit cannot be effectively quantified by traditional *laissez-faire* economics. Allen Hershkowitz supports intervention, saying that it is a public service equal to education and policing. He argues that manufacturers should shoulder more of the burden of waste disposal.

Wrecked automobiles gathered for smelting.

Paul Calcott and Margaret Walls advocate the second option. A deposit refund scheme and a small refuse charge would encourage recycling but not at the expense of fly-tipping. Thomas C. Kinnaman concludes that a landfill tax would force consumers, companies and councils to recycle more.

Most free-market thinkers detest subsidy and intervention because they waste resources. Terry Anderson and Donald Leal think that all recycling programmes should be privately operated, and therefore would only operate if the money saved by recycling exceeds its costs. Daniel K. Benjamin argues that it wastes people's resources and lowers the wealth of a population.

Trade in Recyclates

Certain countries trade in unprocessed recyclates. Some have complained that the ultimate fate of recyclates sold to another country is unknown and they may end up in landfills instead of reprocessed. According to one report, in America, 50–80 percent of computers destined for recycling are actually not recycled. There are reports of illegal-waste imports to China being dismantled and recycled solely for monetary gain, without consideration for workers' health or environmental damage. Although the Chinese government has banned these practices, it has not been able to eradicate them. In 2008, the prices of recyclable waste plummeted before rebounding in 2009. Cardboard averaged about £53/tonne from 2004–2008, dropped to £19/tonne, and then went up to £59/tonne in May 2009. PET plastic averaged about £156/tonne, dropped to £75/tonne and then moved up to £195/tonne in May 2009.

Certain regions have difficulty using or exporting as much of a material as they recycle. This problem is most prevalent with glass: both Britain and the U.S. import large quantities of wine bottled in green glass. Though much of this glass is sent to be recycled, outside the American Midwest there is not enough wine production to use all of the reprocessed material. The extra must be downcycled into building materials or re-inserted into the regular waste stream.

Similarly, the northwestern United States has difficulty finding markets for recycled newspaper, given the large number of pulp mills in the region as well as the proximity to Asian markets. In other areas of the U.S., however, demand for used newsprint has seen wide fluctuation.

In some U.S. states, a program called RecycleBank pays people to recycle, receiving money from local municipalities for the reduction in landfill space which must be purchased. It uses a single stream process in which all material is automatically sorted.

Criticisms and Responses

Much of the difficulty inherent in recycling comes from the fact that most products are not designed with recycling in mind. The concept of sustainable design aims to solve this problem, and was laid out in the book *Cradle to Cradle: Remaking the Way We Make Things* by architect William McDonough and chemist Michael Braungart. They suggest that every product (and all packaging they require) should have a complete "closed-loop" cycle mapped out for each component—a way in which every component will either return to the natural ecosystem through biodegradation or be recycled indefinitely.

Complete recycling is impossible from a practical standpoint. In summary, substitution and recycling strategies only delay the depletion of non-renewable stocks and therefore may buy time in the transition to true or strong sustainability, which ultimately is only guaranteed in an economy based on renewable resources.

—M. H. Huesemann, 2003

While recycling diverts waste from entering directly into landfill sites, current recycling misses the dissipative components. Complete recycling is impracticable as highly dispersed wastes become so diluted that the energy needed for their recovery becomes increasingly excessive. "For example, how will it ever be possible to recycle the numerous chlorinated organic hydrocarbons that have bioaccumulated in animal and human tissues across the globe, the copper dispersed in fungicides, the lead in widely applied paints, or the zinc oxides present in the finely dispersed rubber powder that is abraded from automobile tires?"

As with environmental economics, care must be taken to ensure a complete view of the costs and benefits involved. For example, paperboard packaging for food products is more easily recycled than most plastic, but is heavier to ship and may result in more waste from spoilage.

Energy and Material Flows

The amount of energy saved through recycling depends upon the material being recycled and the type of energy accounting that is used. Correct accounting for this saved energy can be accomplished with life-cycle analysis using real energy values. In addition, exergy, which is a measure of useful energy can be used. In general, it takes far less energy to produce a unit mass of recycled materials than it does to make the same mass of virgin materials.

Some scholars use emergy (spelled with an m) analysis, for example, budgets for the amount of energy of one kind (exergy) that is required to make or transform things into another kind of product or service. Emergy calculations take into account economics which can alter pure physics based results. Using emergy life-cycle analysis researchers have concluded that materials with large refining costs have the greatest potential for high recycle benefits. Moreover, the highest emergy efficiency accrues from systems geared toward material recycling, where materials are engineered to recycle back into their original form and purpose, followed by adaptive reuse systems where the materials are recycled into a different kind of product, and then by-product reuse systems where parts of the products are used to make an entirely different product.

The Energy Information Administration (EIA) states on its website that "a paper mill uses 40 percent less energy to make paper from recycled paper than it does to make paper from fresh lumber." Some critics argue that it takes more energy to produce recycled products than it does to dispose of them in traditional landfill methods, since the curbside collection of recyclables often requires a second waste truck. However, recycling proponents point out that a second timber or logging truck is eliminated when paper is collected for recycling, so the net energy consumption is the same. An Emergy life-cycle analysis on recycling revealed that fly ash, aluminum, recycled concrete aggregate, recycled plastic, and steel yield higher efficiency ratios, whereas the recycling of lumber generates the lowest recycle benefit ratio. Hence, the specific nature of the recycling process, the methods used to analyse the process, and the products involved affect the energy savings budgets.

It is difficult to determine the amount of energy consumed or produced in waste disposal processes in broader ecological terms, where causal relations dissipate into complex networks of material and energy flow. For example, "cities do not follow all the strategies of ecosystem development. Biogeochemical paths become fairly straight relative to wild ecosystems, with very reduced recycling, resulting in large flows of waste and low total energy efficiencies. By contrast, in wild ecosystems, one population's wastes are another population's resources, and succession results in efficient exploitation of available resources. However, even modernized cities may still be in the earliest stages of a succession that may take centuries or millennia to complete." How much energy is used in recycling also depends on the type of material being recycled and the process used to do so. Aluminium is generally agreed to use far less energy when recycled rather than being produced from scratch. The EPA states that "recycling aluminum cans, for example, saves 95 percent of the energy required to make the same amount of aluminum from its virgin source, bauxite." In 2009 more than half of all aluminium cans produced came from recycled aluminium.

Every year, millions of tons of materials are being exploited from the earth's crust, and processed into consumer and capital goods. After decades to centuries, most of these materials are "lost". With the exception of some pieces of art or religious relics, they are no longer engaged in the consumption process. Where are they? Recycling is only an intermediate solution for such materials, although it does prolong the residence time in the anthroposphere. For thermodynamic reasons, however, recycling cannot prevent the final need for an ultimate sink.

—P. H. Brunner

Economist Steven Landsburg has suggested that the sole benefit of reducing landfill space is trumped by the energy needed and resulting pollution from the recycling process. Others, however, have calculated through life-cycle assessment that producing recycled paper uses less energy and water than harvesting, pulping, processing, and transporting virgin trees. When less recycled paper is used, additional energy is needed to create and maintain farmed forests until these forests are as self-sustainable as virgin forests.

Other studies have shown that recycling in itself is inefficient to perform the "decoupling" of economic development from the depletion of non-renewable raw materials that is necessary for sustainable development. The international transportation or recycle material flows through "... different trade networks of the three countries result in different flows, decay rates, and potential recycling returns." As global consumption of a natural resources grows, its depletion is inevitable. The best recycling can do is to delay, complete closure of material loops to achieve 100 percent recycling of nonrenewables is impossible as micro-trace materials dissipate into the environment causing severe damage to the planet's ecosystems. Historically, this was identified as the metabolic rift by Karl Marx, who identified the unequal exchange rate between energy and nutrients flowing from rural areas to feed urban cities that create effluent wastes degrading the planet's ecological capital, such as loss in soil nutrient production. Energy conservation also leads to what is known as Jevon's paradox, where improvements in energy efficiency lowers the cost of production and leads to a rebound effect where rates of consumption and economic growth increases.

A shop in New York only sells items recycled from demolished buildings.

Costs

The amount of money actually saved through recycling depends on the efficiency of the recycling program used to do it. The Institute for Local Self-Reliance argues that the cost of recycling depends on various factors, such as landfill fees and the amount of disposal that the community recycles. It states that communities begin to save money when they treat recycling as a replacement for their traditional waste system rather than an add-on to it and by "redesigning their collection schedules and/or trucks."

In some cases, the cost of recyclable materials also exceeds the cost of raw materials. Virgin plastic resin costs 40 percent less than recycled resin. Additionally, a United States Environmental Protection Agency (EPA) study that tracked the price of clear glass from July 15 to August 2, 1991, found that the average cost per ton ranged from $40 to $60, while a USGS report shows that the cost per ton of raw silica sand from years 1993 to 1997 fell between $17.33 and $18.10.

In 1996 and 2015 articles for *The New York Times*, John Tierney argued that it costs more money to recycle the trash of New York City than it does to dispose of it in a landfill. Tierney argued that the recycling process employs people to do the additional waste disposal, sorting, inspecting, and many fees are often charged because the processing costs used to make the end product are often more than the profit from its sale. Tierney also referenced a study conducted by the Solid Waste Association of North America (SWANA) that found in the six communities involved in the study, "all but one of the curbside recycling programs, and all the composting operations and waste-to-energy incinerators, increased the cost of waste disposal."

Tierney also points out that "the prices paid for scrap materials are a measure of their environmental value as recyclables. Scrap aluminum fetches a high price because recycling it consumes so much less energy than manufacturing new aluminum."

However, comparing the market cost of recyclable material with the cost of new raw materials ignores economic externalities—the costs that are currently not counted by the market. Creating a new piece of plastic, for instance, may cause more pollution and be less sustainable than recycling a similar piece of plastic, but these factors will not be counted in market cost. A life cycle assessment can be used to determine the levels of externalities and decide whether the recycling may be worthwhile despite unfavorable market costs. Alternatively, legal means (such as a carbon tax) can be used to bring externalities into the market, so that the market cost of the material becomes close to the true cost.

In a 2007 article, Michael Munger, chairman of political science at Duke University, wrote that "if recycling is more expensive than using new materials, it can't possibly be efficient.... There is a simple test for determining whether something is a resource... or just garbage... If someone will pay you for the item, it's a resource.... But if you have to pay someone to take the item away,... then the item is garbage."

In a 2002 article for The Heartland Institute, Jerry Taylor, director of natural resource studies at the Cato Institute, wrote, "If it costs X to deliver newly manufactured plastic to the market, for example, but it costs 10X to deliver reused plastic to the market, we can conclude the resources required to recycle plastic are 10 times more scarce than the resources required to make plastic from scratch. And because recycling is supposed to be about the conservation of resources, mandating recycling under those circumstances will do more harm than good."

Working Conditions

The recycling of waste electrical and electronic equipment in India and China generates a significant amount of pollution. Informal recycling in an underground economy of these countries has generated an environmental and health disaster. High levels of lead (Pb), polybrominated diphenylethers (PBDEs), polychlorinated dioxins and furans, as well as polybrominated dioxins and furans (PCDD/Fs and PBDD/Fs) concentrated in the air, bottom ash, dust, soil, water and sediments in areas surrounding recycling sites. Critics also argue that while recycling may create jobs, they are often jobs with low wages and terrible working conditions. These jobs are sometimes considered to be make-work jobs that don't produce as much as the cost of wages to pay for those jobs. In areas without many environmental regulations and/or worker protections, jobs involved in recycling such as ship breaking can result in deplorable conditions for both workers and the surrounding communities.

People in Brazil who earn their living by collecting and sorting garbage and selling them for recycling.

Environmental Impact

Economist Steven Landsburg, author of a paper entitled "Why I Am Not an Environmentalist," claimed that paper recycling actually reduces tree populations. He argues that because paper companies have incentives to replenish their forests, large demands for paper lead to large forests, while reduced demand for paper leads to fewer "farmed" forests.

When foresting companies cut down trees, more are planted in their place. Most paper comes from pulp forests grown specifically for paper production. Many environmentalists point out, however, that "farmed" forests are inferior to virgin forests in several ways.

Farmed forests are not able to fix the soil as quickly as virgin forests, causing widespread soil erosion and often requiring large amounts of fertilizer to maintain while containing little tree and wild-life biodiversity compared to virgin forests. Also, the new trees planted are not as big as the trees that were cut down, and the argument that there will be "more trees" is not compelling to forestry advocates when they are counting saplings.

In particular, wood from tropical rainforests is rarely harvested for paper because of their heterogeneity. According to the United Nations Framework Convention on Climate Change secretariat, the overwhelming direct cause of deforestation is subsistence farming (48% of deforestation) and commercial agriculture (32%), which is linked to food, not paper production.

Possible Income Loss and Social Costs

In some countries, recycling is performed by the entrepreneurial poor such as the karung guni, zabbaleen, the rag-and-bone man, waste picker, and junk man. With the creation of large recycling organizations that may be profitable, either by law or economies of scale, the poor are more likely to be driven out of the recycling and the remanufacturing market. To compensate for this loss of income, a society may need to create additional forms of societal programs to help support the poor. Like the parable of the broken window, there is a net loss to the poor and possibly the whole of a society to make recycling artificially profitable e.g. through the law. However, in Brazil and Argentina, waste pickers/informal recyclers work alongside the authorities, in fully or semi-funded cooperatives, allowing informal recycling to be legitimized as a paid public sector job.

Because the social support of a country is likely to be less than the loss of income to the poor undertaking recycling, there is a greater chance the poor will come in conflict with the large recycling organizations. This means fewer people can decide if certain waste is more economically reusable in its current form rather than being reprocessed. Contrasted to the recycling poor, the efficiency of their recycling may actually be higher for some materials because individuals have greater control over what is considered "waste."

One labor-intensive underused waste is electronic and computer waste. Because this waste may still be functional and wanted mostly by those on lower incomes, who may sell or use it at a greater efficiency than large recyclers.

Some recycling advocates believe that laissez-faire individual-based recycling does not cover all of society's recycling needs. Thus, it does not negate the need for an organized recycling program. Local government can consider the activities of the recycling poor as contributing to property blight.

Public Participation Rates

Changes that have been demonstrated to increase recycling rates include:

- Single-stream recycling
- Pay as you throw fees for trash

"Between 1960 and 2000, the world production of plastic resins increased 25-fold, while recovery of the material remained below 5 percent." Many studies have addressed recycling behaviour and strategies to encourage community involvement in recycling programmes. It has been argued that recycling behaviour is not natural because it requires a focus and appreciation for long-term planning, whereas humans have evolved to be sensitive to short-term survival goals; and that to overcome this innate predisposition, the best solution would be to use social pressure to compel participation in recycling programmes. However, recent studies have concluded that social pressure is unviable in this context. One reason for this is that social pressure functions well in small group sizes of 50 to 150 individuals (common to nomadic hunter–gatherer peoples) but not in communities numbering in the millions, as we see today. Another reason is that individual recycling does not take place in the public view.

In a study done by social psychologist Shawn Burn, it was found that personal contact with individuals within a neighborhood is the most effective way to increase recycling within a community. In his study, he had 10 block leaders talk to their neighbors and persuade them to recycle. A comparison group was sent fliers promoting recycling. It was found that the neighbors that were personally contacted by their block leaders recycled much more than the group without personal contact. As a result of this study, Shawn Burn believes that personal contact within a small group of people is an important factor in encouraging recycling. Another study done by Stuart Oskamp examines the effect of neighbors and friends on recycling. It was found in his studies that people who had friends and neighbors that recycled were much more likely to also recycle than those who didn't have friends and neighbors that recycled.

Many schools have created recycling awareness clubs in order to give young students an insight on recycling. These schools believe that the clubs actually encourage students to not only recycle at school, but at home as well.

Plastic Recycling

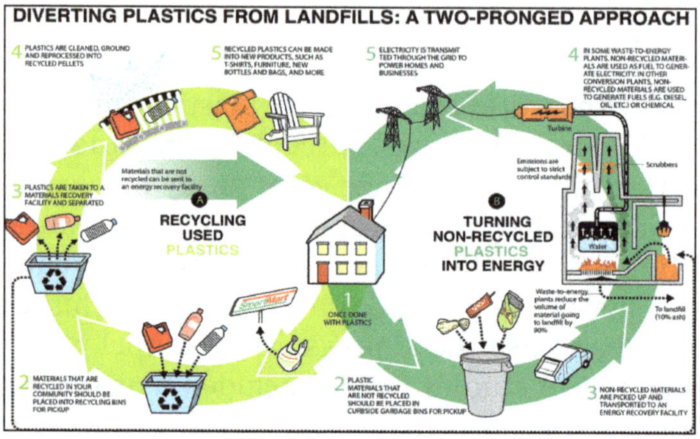

Plastic recycling is the process of recovering scrap or waste plastic and reprocessing the material into useful products. Since plastic is non-biodegradable, recycling it is a part of global efforts to reduce plastic in the waste stream, especially the approximately eight million metric tonnes of waste plastic that enter the Earth's ocean every year. This helps to reduce the high rates of plastic pollution.

Plastic recycling includes taking any type of plastic sorting it into different polymers and then chipping it and then melting it down into pellets after this stage it can then be used to make items of any kind such as plastic chairs and tables. Soft Plastics are also recycled such as polyethylene film and bags. This closed-loop operation has taken place since the 1970s and has made the production of some plastic products amongst the most efficient operations today.

Compared with lucrative recycling of metal, and similar to the low value of glass, plastic polymers recycling is often more challenging because of low density and low value. There are also numerous technical hurdles to overcome when recycling plastic.

A macro molecule interacts with its environment along its entire length, so total energy involved in mixing it is largely due to the product side stoichiometry. Heating alone is not enough to dissolve such a large molecule, so plastics must often be of nearly identical composition to mix efficiently.

When different types of plastics are melted together, they tend to phase-separate, like oil and water, and set in these layers. The phase boundaries cause structural weakness in the resulting material, meaning that polymer blends are useful in only limited applications.

Another barrier to recycling is the widespread use of dyes, fillers, and other additives in plastics. The polymer is generally too viscous to economically remove fillers, and would be damaged by many of the processes that could cheaply remove the added dyes. Additives are less widely used in beverage containers and plastic bags, allowing them to be recycled more often. Yet another barrier to removing large quantities of plastic from the waste stream and landfills is the fact that many common but small plastic items lack the universal triangle recycling symbol and accompanying number. An example is the billions of plastic utensils commonly distributed at fast food restaurants or sold for use at picnics.

The percentage of plastic that can be fully recycled, rather than downcycled or go to waste can be increased when manufacturers of packaged goods minimize mixing of packaging materials and eliminate contaminants. The Association of Plastics Recyclers have issued a Design Guide for Recyclability.

The use of biodegradable plastics is increasing.

Processes

Before recycling, most plastics are sorted according to their resin type. In the past, plastic reclaimers used the resin identification code (RIC), a method of categorization of polymer types, which was developed by the Society of the Plastics Industry in 1988. polyethylene terephthalate, commonly referred to as PET, for instance, has a resin code of 1. Most plastic reclaimers do not rely on the RIC now; they use automatic sort systems to identify the resin. Ranging from manual sorting and picking of plastic materials; to mechanized automation processes that involve shredding, sieving, separation by rates of density i.e. air, liquid, or magnetic, and complex spectrophotometric distribution technologies e.g. UV/VIS, NIR, Laser, etc. Some plastic products are also separated by color before they are recycled. The plastic recyclables are then shredded. These shredded fragments then undergo processes to eliminate impurities like paper labels. This material is melted and often extruded into the form of pellets which are then used to manufacture other products.

Thermal Depolymerization

Another process involves the conversion of assorted polymers into petroleum by a much less precise thermal depolymerization process. Such a process would be able to accept almost any polymer or mix of polymers, including thermoset materials such as vulcanized rubber tire separation of wastes and the biopolymers in feathers and other agricultural waste. Like natural petroleum, the chemicals produced can be made into fuels as well as polymers. A pilot plant of this type exists in Carthage, Missouri, United States, using turkey waste as input material. Gasification is a similar process, but is not technically recycling, since polymers are not likely to become the result.

Heat Compression

Yet another process that is gaining ground with startup companies (especially in Australia, United States and Japan) is heat compression. The heat compression process takes all unsorted, cleaned plastic in all forms, from soft plastic bags to hard industrial waste, and mixes the load in tumblers (large rotating drums resembling giant clothes dryers). The most obvious benefit to this method is the fact that all plastic is recyclable, not just matching forms. However, criticism rises from the energy costs of rotating the drums, and heating the post-melt pipes.

Distributed Recycling

For some waste plastics, recent technical devices called recyclebots enable a form of distributed recycling. Preliminary life-cycle analysis(LCA) indicates that such distributed recycling of HDPE to make filament of 3-D printers in rural regions is energetically favorable to either using virgin resin or conventional recycling processes because of reductions in transportation energy.

Other Processes

A process has also been developed in which many kinds of plastic can be used as a carbon source in the recycling of scrap steel. There is also a possibility of mixed recycling of different plastics, which does not require their separation. It is called Compatibilization and requires use of special chemical bridging agents compatibilizers. It can help to keep the quality of recycled material and to skip often expensive and inefficient preliminary scanning of waste plastics streams and their separation/purification.

Applications

PET

Post-consumer polyethylene terephthalate (PET or PETE) containers are sorted into different colour fractions, and baled for onward sale. PET recyclers further sort the baled bottles and they are washed and flaked (or flaked and then washed). Non-PET fractions such as caps and labels are removed during this process. The clean flake is dried. Further treatment can take place e.g. melt filtering and pelletising or various treatments to produce food-contact-approved recycled PET (RPET).

RPET has been widely used to produce polyester fibres. This sorted post-consumer PET waste is crushed, chopped into flakes, pressed into bales, and offered for sale.

One use for this recycled PET that has recently started to become popular is to create fabrics to be used in the clothing industry. The fabrics are created by spinning the PET flakes into thread and yarn. This is done just as easily as creating polyester from brand new PET. The recycled PET thread or yarn can be used either alone or together with other fibers to create a very wide variety of fabrics. Traditionally these fabrics are used to create strong, durable, rough, products, such as jackets, coat, shoes, bags, hats, and accessories since they are usually too rough for direct skin contact and can cause irritation. However, these types of fabrics have become more popular as a result of the public's growing awareness of environmental issues. Numerous fabric and clothing manufacturers have capitalized on this trend.

Other major outlets for RPET are new containers (food-contact or non-food-contact) produced either by (injection stretch blow) moulding into bottles and jars or by thermoforming APET sheet to produce clam shells, blister packs and collation trays. These applications used 46% of all RPET produced in Europe in 2010. Other applications, such as strapping tape, injection-moulded engineering components and even building materials account for 13% of the 2010 RPET production.

In the United States the recycling rate for PET packaging was 31.2% in 2013, according to a report from The National Association for PET Container Resources (NAPCOR) and The Association of Postconsumer Plastic Recyclers (APR). A total of 1,798 million pounds was collected and 475 million pounds of recycled PET used out of a total of 5,764 million pounds of PET bottles.

HDPE

Plastic # 2, high-density polyethylene (HDPE) is a commonly recycled plastic. It is typically downcycled into plastic lumber, tables, roadside curbs, benches, truck cargo liners, trash receptacles, stationery (e.g. rulers) and other durable plastic products and is usually in demand.

PS

Most polystyrene products are currently not recycled due to the lack of incentive to invest in the compactors and logistical systems required. As a result, manufacturers cannot obtain sufficient scrap. Expanded polystyrene (EPS) scrap can easily be added to products such as EPS insulation sheets and other EPS materials for construction applications. When it is not used to make more EPS, foam scrap can be turned into clothes hangers, park benches, flower pots, toys, rulers, stapler bodies, seedling containers, picture frames, and architectural molding from recycled PS.

The resin identification code symbol for polystyrene.

Recycled EPS is also used in many metal casting operations. Rastra is made from EPS that is combined with cement to be used as an insulating amendment in the making of concrete foundations and walls. Since 1993, American manufacturers have produced insulating concrete forms made with approximately 80% recycled EPS.

Other Plastics

The white plastic polystyrene foam peanuts used as packing material are often accepted by shipping stores for reuse.

Successful trials in Israel have shown that plastic films recovered from mixed municipal waste streams can be recycled into useful household products such as buckets.

Similarly, agricultural plastics such as mulch film, drip tape and silage bags are being diverted from the waste stream and successfully recycled into much larger products for industrial applications such as plastic composite railroad ties. Historically, these agricultural plastics have primarily been either landfilled or burned on-site in the fields of individual farms.

CNN reports that Dr. S. Madhu of the Kerala Highway Research Institute, India, has formulated a road surface that includes recycled plastic: aggregate, bitumen (asphalt) with plastic that has been shredded and melted at a temperature below 220 degrees C (428 °F) to avoid pollution. This road surface is claimed to be very durable and monsoon rain resistant. The plastic is sorted by hand, which is economical in India. The test road used 60 kg of plastic for an approximately 500m-long, 8m-wide, two-lane road. The process chops thin-film road-waste into a light fluff of tiny flakes that hot-mix plants can uniformly introduce into viscous bitumen with a customized dosing machine. Tests at both Bangalore and the Indian Road Research Centre indicate that roads built using this 'KK process' will have longer useful lives and better resistance to cold, heat, cracking, and rutting, by a factor of three.

Recycling Rates

The quantity of post-consumer plastics recycled has increased every year since at least 1990, but rates lag far behind those of other items, such as newspaper (about 80%) and corrugated fiberboard (about 70%). Overall, U.S. post-consumer plastic waste for 2008 was estimated at 33.6 million tons; 2.2 million tons (6.5%) were recycled and 2.6 million tons (7.7%) were burned for energy; 28.9 million tons, or 85.5%, were discarded in landfills.

Economic and Energy Potential

In 2008, the price of PET dropped from $370/ton in the US to $20 in November. PET prices had returned to their long-term averages by May 2009.

Recycling one ton of plastic can save 5,774 kWh of energy, 98,000,000 btus of energy, 1,000-2,000 gallons of gasoline, 685 gallons of oil, 30 cubic yards of landfill space, 48,000 gallons of water.

Consumer Education

United Kingdom

In the UK, the amount of post-consumer plastic being recycled is relatively low, due in part to a lack of recycling facilities.

The Plastics 2020 Challenge was founded in 2009 by the plastics industry with the aim of engaging the British public in a nationwide debate about the use, reuse and disposal of plastics, and hosts a series of online debates on its website framed around the waste hierarchy.

There is a facility in Worksop capable of recycling 60–80 thousand metric tonnes a year.

In Northern Ireland, the rate of recycling is relatively low at only 37.4%. However, emerging technologies are helping to increase the recycling rates of items previously landfilled e.g. mixed hard plastics.

Plastic Identification Code

Five groups of plastic polymers, each with specific properties, are used worldwide for packaging applications. Each group of plastic polymer can be identi-fied by its Plastic Identification code (PIC), usually a number or a letter abbreviation. For instance, Low-Density Polyethylene can be identified by the number "4" or the letters "LDPE". The PIC appears inside a three-chasing-arrow recycling symbol. The symbol is used to indicate whether the plastic can be recycled into new products.

The PIC was introduced by the Society of the Plastics Industry, Inc., to provide a uniform system for the identification of various polymer types and to help recycling companies separate various plastics for reprocessing. Manufacturers of plastic products are required to use PIC labels in some countries/regions and can voluntarily mark their products with the PIC where there are no requirements. Consumers can identify the plastic types based on the codes usually found at the base or at the side of the plastic products, including food/chemical packaging and containers. The PIC is usually not present on packaging films, since it is not practical to collect and recycle most of this type of waste.

United States

Low national plastic recycling rates have been due to the complexity of sorting and processing, unfavorable economics, and consumer confusion about which plastics can actually be recycled. Part of the confusion has been due to the use of the resin identification code which is not on all plastic parts but just a subset that includes the recycling symbol as part of its design. The resin identification code is stamped or printed on the bottom of containers and surrounded by a triangle of arrows. The intent of these symbols was to make it easier to identify the type of plastics used to make a particular container and to indicate that the plastic is potentially recyclable. The question that remains is which types of plastics can be recycled by your local recycling center. In many communities, not all types of plastics are accepted for sidewalk recycling collection programs due to the high processing costs and complexity of the equipment required to recycle certain materials. There is also sometimes a seemingly low demand for the recycled product depending on a recycling center's proximity to entities seeking recycled materials. Another major barrier is that the cost to recycle certain materials and the corresponding market price for those materials sometimes does not present any opportunity for profit. The best example of this is polystyrene (commonly called styrofoam), although some communities, like Brookline, MA, are moving toward banning the distribution of polystyrene containers by local food and coffee businesses.

Computer Recycling

Computer recycling, electronic recycling or e-waste recycling is the disassembly and separation of components and raw materials of waste electronics. Although the procedures of re-use, donation and repair are not strictly recycling, they are other common sustainable ways to dispose of IT waste.

Computer monitors are typically packed into low stacks on wooden pallets for recycling and then shrink-wrapped.

In 2009, 38% of computers and a quarter of total electronic waste was recycled in the United States, 5% and 3% up from 3 years prior respectively. Since its inception in the early 1990s, more and more devices are recycled worldwide due to increased awareness and investment. Electronic recycling occurs primarily in order to recover valuable rare earth metals and precious metals, which are in short supply, as well as plastics and metals. These are resold or used in new devices after purification, in effect creating a circular economy.

Recycling is considered environmentally friendly because it prevents hazardous waste, including heavy metals and carcinogens, from entering the atmosphere, landfill or waterways. While electronics consist a small fraction of total waste generated, they are far more dangerous. There is stringent legislation designed to enforce and encourage the sustainable disposal of appliances, the most notable being the Waste Electrical and Electronic Equipment Directive of the European Union and the United States National Computer Recycling Act.

Opponents argue that recycling is expensive and ineffective, that it does not safeguard data and that it stifles innovation. It is also criticised for exporting, often illegally, large volumes of toxic waste to countries such as India, China and Nigeria for crude manual disassembly by workers who have little regard for the risk to themselves or the environment.

Reasons for Recycling

Obsolete computers and old electronics are valuable sources for secondary raw materials if recycled; otherwise, these devices are a source of toxins and carcinogens. Rapid technology change, low initial cost, and planned obsolescence have resulted in a fast-growing surplus of computers and other electronic components around the globe. Technical solutions are available, but in most cases a legal framework, collection system, logistics, and other services need to be implemented before applying a technical solution. The U.S. Environmental Protection Agency, estimates 30 to 40 million surplus PCs, classified as "hazardous household waste", would be ready for end-of-life management in the next few years. The U.S. National Safety Council estimates that 75% of all personal computers ever sold are now surplus electronics.

In 2007, the United States Environmental Protection Agency (EPA) stated that more than 63 million computers in the U.S. were traded in for replacements or discarded. Today, 15% of electronic devices and equipment are recycled in the United States. Most electronic waste is sent to landfills or incinerated, which releases materials such as lead, mercury, or cadmium into the soil, groundwater, and atmosphere, thus having a negative impact on the environment.

Many materials used in computer hardware can be recovered by recycling for use in future production. Reuse of tin, silicon, iron, aluminium, and a variety of plastics that are present in bulk in computers or other electronics can reduce the costs of constructing new systems. Components frequently contain lead, copper, gold and other valuable materials suitable for reclamation.

Computer components contain many toxic substances, like dioxins, polychlorinated biphenyls (PCBs), cadmium, chromium, radioactive isotopes and mercury. A typical computer monitor may contain more than 6% lead by weight, much of which is in the lead glass of the cathode ray tube (CRT). A typical 15 inch (38 cm) computer monitor may contain 1.5 pounds (1 kg) of lead but other monitors have been estimated to have up to 8 pounds (4 kg) of lead. Circuit boards contain considerable quantities of lead-tin solders that are more likely to leach into groundwater or create air pollution due to incineration. The processing (e.g. incineration and acid treatments) required to reclaim these precious substances may release, generate, or synthesize toxic by-products.

Export of waste to countries with lower environmental standards is a major concern. The Basel Convention includes hazardous wastes such as, but not limited to, CRT screens as an item that may not be exported transcontinentally without prior consent of both the country exporting and receiving the waste. Companies may find it cost-effective in the short term to sell outdated computers to less developed countries with lax regulations. It is commonly believed that a majority of surplus laptops are routed to developing nations as "dumping grounds for e-waste". The high value of working and reusable laptops, computers, and components (e.g. RAM) can help pay the cost of transportation for many worthless "commodities". The laws governing the exportation of waste electronics are put in place to stop "recycling companies" in developed countries from shipping their waste to 3rd world countries as working devices; they are never working devices. The 3rd world workers scavenge specific items with selling value and throw the rest away to rot and become a health hazard in their own backyard.

Regulations

In Switzerland, the first electronic waste recycling system was implemented in 1991, beginning with collection of old refrigerators; over the years, all other electric and electronic devices were gradually added to the system. The established producer responsibility organization is SWICO, mainly handling information, communication, and organization technology. The European Union implemented a similar system in February 2003, under the Waste Electrical and Electronic Equipment Directive (WEEE Directive, 2002/96/EC).

Pan European adoption of the Legislation was slow on take-up, with Italy and the United Kingdom being the final member states to pass it into law. The success of the WEEE directive has varied significantly from state to state, with collection rates varying between 13 kilograms per capita per annum to as little as 1 kg per capita per annum. Computers & electronic wastes collected from households within Europe are treated under the WEEE directive via Producer Compliance Schemes (whereby manufacturers of Electronics pay into a scheme that funds its recovery from household waste recycling centres (HWRCs)) and nominated Waste Treatment Facilities (known as Obligated WEEE).

An abandoned Taxan monitor.

Europe

However, recycling of ex corporate Computer Hardware and associated electronic equipment falls outside the Producer Compliance Scheme (Known as non-obligated). In the UK, Waste or obsolete corporate related computer hardware is treated via third party Authorized Treatment Facilities, who normally impose a charge for its collection and treatment.

United States

Federal

The United States Congress considers a number of electronic waste bills, like the National Computer Recycling Act introduced by Congressman Mike Thompson (D-CA). The main federal law governing solid waste is the Resource Conservation and Recovery Act of 1976. It covers only CRTs, though state regulations may differ. There are also separate laws concerning battery disposal. On March 25, 2009, the House Science and Technology Committee approved funding for research on reducing electronic waste and mitigating environmental impact, regarded by sponsor Ralph Hall (R-TX) as the first federal bill to directly address electronic waste.

State

Many states have introduced legislation concerning recycling and reuse of computers or computer parts or other electronics. Most American computer recycling legislations address it from within the larger electronic waste issue.

In 2001, Arkansas enacted the Arkansas Computer and Electronic Solid Waste Management Act, which requires that state agencies manage and sell surplus computer equipment, establishes a computer and electronics recycling fund, and authorizes the Department of Environmental Quality to regulate and/or ban the disposal of computer and electronic equipment in Arkansas landfills.

The recently passed Electronic Device Recycling Research and Development Act distributes grants to universities, government labs and private industries for research in developing projects in line with e-waste recycling and refurbishment.

Asia

In Japan, sellers and manufacturers of certain electronics (such as televisions and air conditioners) are required to recycle them. However, no legislation exists to cover the recycling of computer or cellphone related wastes.

It is required in South Korea and Taiwan that sellers and manufacturers of electronics be responsible for recycling 75% of their used products.

According to a report by UNEP titled, “Recycling - from E-Waste to Resources,” the amount of e-waste being produced - including mobile phones and computers - could rise by as much as 500 percent over the next decade in some countries, such as India.

Electronic waste is often exported to developing countries.

4.5-volt, D, C, AA, AAA, AAAA, A23, 9-volt, CR2032 and LR44 cells are all recyclable in most countries.

One theory is that increased regulation of electronic waste and concern over the environmental harm in mature economies creates an economic disincentive to remove residues prior to export. Critics of trade in used electronics maintain that it is too easy for brokers calling themselves recyclers to export unscreened electronic waste to developing countries, such as China, India and parts of Africa, thus avoiding the expense of removing items like bad cathode ray tubes (the processing of which is expensive and difficult). The developing countries are becoming big dump yards of e-waste. Proponents of international trade point to the success of fair trade programs in other industries, where cooperation has led creation of sustainable jobs, and can bring affordable technology in countries where repair and reuse rates are higher.

Organizations like A2Z Group (Company Website) have stepped in to own up the responsibility to collect and recycle e-waste at various locations in India.

South Africa

Thanks to the National Environmental Management Act 1998 and National Environmental Management Waste Act 2008, any person in any position causing harm to the environment and failing to comply with the Waste Act could be fined R10 Million or put into jail or receive both penalties for their transgressions.

Recycling Methods

Computers being collected for recycling at a pickup event in Olympia, Washington, United States.

Consumer Recycling

Consumer recycling options consists of sale, donating computers directly to organizations in need, sending devices directly back to their original manufacturers, or getting components to a convenient recycler or refurbisher.

Scrapping/Recycling

The rising price of precious metals — coupled with the high rate of unemployment during the Great Recession — has led to a larger number of amateur "for profit" electronics recyclers. Computer parts, for example, are stripped of their most valuable components and sold for scrap. Metals like copper, aluminum, lead, gold and palladium are recovered from computers, televisions and more.

In the recycling process, TVs, monitors, mobile phones and computers are typically tested for reuse and repaired. If broken, they may be disassembled for parts still having high value if labour is cheap enough. Other e-waste is shredded to roughly 100 mm pieces and manually checked to separate out toxic batteries and capacitors which contain poisonous metals. The remaining pieces are further shredded to ~10 mm and passed under a magnet to remove ferrous metals. An eddy current ejects non-ferrous metals, which are sorted by density either by a centrifuge or vibrating plates. Precious metals can be dissolved in acid, sorted, and smelted into ingots. The remaining glass and plastic fractions are separated by density and sold to re-processors. TVs and monitors must be manually disassembled to remove either toxic lead in CRTs or the mercury in flat screens.

Bulk laptops at a recycling affiliate, broken down into Dell, Gateway Computers, Hewlett-Packard, Sony, and other.

Corporate Recycling

Businesses seeking a cost-effective way to recycle large amounts of computer equipment responsibly face a more complicated process.

Businesses also have the options of sale or contacting the Original Equipment Manufacturers (OEMs) and arranging recycling options.

Some companies pick up unwanted equipment from businesses, wipe the data clean from the systems, and provide an estimate of the product's remaining value. For unwanted items that still have value, these firms buy the excess IT hardware and sell refurbished products to those seeking more affordable options than buying new.

Companies that specialize in data protection and green disposal processes dispose of both data and used equipment, while employing strict procedures to help improve the environment. Professional IT Asset Disposition (ITAD) firms specialize in corporate computer disposal and recycling services in compliance with local laws and regulations and also offer secure data elimination services that comply with Data remanence standards including National Institute of Standards and Technology.

Corporations face risks both for incompletely destroyed data and for improperly disposed computers. In the UK, some recycling companies use a specialized WEEE-registered contractor to dispose IT equipment and electrical appliances, who disposes it safely and legally. In America, companies are liable for compliance with regulations even if the recycling process is outsourced under the Resource Conservation and Recovery Act. Companies can mitigate these risks by requiring waivers of liability, audit trails, certificates of data destruction, signed confidentiality agreements, and random audits of information security. The National Association of Information Destruction is an international trade association for data destruction providers.

Sale

Online auctions are an alternative for consumers willing to resell for cash less fees, in a complicated, self-managed, competitive environment where paid listings might not sell. Online classified ads can be similarly risky due to forgery scams and uncertainty.

Take Back

When researching computer companies before a computer purchase, consumers can find out if they offer recycling services. Most major computer manufacturers offer some form of recycling. At the user's request they may mail in their old computers, or arrange for pickup from the manufacturer.

Hewlett-Packard also offers free recycling, but only one of its "national" recycling programs is available nationally, rather than in one or two specific states. Hewlett-Packard also offers to pick up any computer product of any brand for a fee, and to offer a coupon against the purchase of future computers or components; it was the largest computer recycler in America in 2003, and it has recycled over 750,000,000 pounds (340,000,000 kg) of electronic waste globally since 1995. It encourages the shared approach of collection points for consumers and recyclers to meet.

Exchange

Manufacturers often offer a free replacement service when purchasing a new PC. Dell Computers and Apple Inc. take back old products when one buys a new one. Both refurbish and resell their own computers with a one-year warranty.

Many companies purchase and recycle all brands of working and broken laptops and notebook computers from individuals and corporations. Building a market for recycling of desktop computers has proven more difficult than exchange programs for laptops, smartphones and other smaller electronics. A basic business model is to provide a seller an instant online quote based on laptop characteristics, then to send a shipping label and prepaid box to the seller, to erase, reformat, and process the laptop, and to pay rapidly by cheque. A majority of these companies are also generalized electronic waste recyclers as well; organizations that recycle computers exclusively include Cash For Laptops, a laptop refurbisher in Nevada that claims to be the first to buy laptops online, in 2001.

Donations/Nonprofits

With the constant rising costs due to inflation, many families or schools do not have the sufficient funds available for computers to be utilized along with education standards. Families also impacted by disaster suffer as well due to the financial impact of the situation they have incurred. Many nonprofit organizations, such as InterConnection.org, can be found locally as well as around the web and give detailed descriptions as to what methods are used for dissemination and detailed instructions on how to donate. The impact can be seen locally and globally, affecting thousands of those in need. In Canada non profit organizations engaged in computer recycling, such as The Electronic Recycling Association Calgary, Edmonton,

Vancouver, Winnipeg, Toronto, Montreal, Computers for Schools Canada wide, are very active in collecting and refurbishing computers and laptops to help the non profit and charitable sectors and schools.

Junkyard Computing

The term *junkyard computing* is a colloquial expression for using old or inferior hardware to fulfill computational tasks while handling reliability and availability on software level. It utilizes abstraction of computational resources via software, allowing hardware replacement at very low effort. Ease of replacement is hereby a corner point since hardware failures are expected at any time due to the condition of the underlying infrastructure. This paradigm became more widely used with the introduction of cluster orchestration software like Kubernetes or Apache Mesos, since large monolithic applications require reliability and availability on machine level whereas this kind of software is fault tolerant by design. Those orchestration tools also introduced fairly fast set-up processes allowing to use junkyard computing economically and even making this pattern applicable in the first place. Further use cases were introduced when continuous delivery was getting more widely accepted. Infrastructure to execute tests and static code analysis was needed which requires as much performance as possible while being extremely cost effective. From an economical and technological perspective, junkyard computing is only practicable for a small amount of users or companies. It already requires a descend amount of physical machines to compensate hardware failures while maintaining the required reliability and availability. This implies a direct need for a matching underling infrastructure to house all the computers and servers. Scaling this paradigm is also quiet limited due to the increasing importance of factors like power efficiency and maintenance efforts, making this kind of computing perfect for mid-sized applications.

History

Although consumer electronics such as the radio have been popular since the 1920s, recycling was almost unheard of until the early 1990s. At the end of the 1970s the accelerating pace of domestic consumer electronics drastically shortened the lifespan of electronics such as TVs, VCRs and audio. New innovations appeared more quickly, making older equipment considered obsolete. Increased complexity and sophistication of manufacture made local repair more difficult. The retail market shifted gradually, but substantially from a few high-value items that were cherished for years and repaired when necessary, to short-lived items that were rapidly replaced owing to wear or simply fashion, and discarded rather than repaired. This was particularly evident in computing, highlighted by Moore's Law. In 1988 two severe incidents highlighted the approaching e-waste crisis. The cargo barge Khian Sea, was loaded with more than 14,000 tons of toxic ash from Pennsylvania which had been refused acceptance in New Jersey and the

Caribbean. After sailing for 16 months, all the waste was dumped as "topsoil fertiliser" in Haiti and in the Bay of Bengal by November 1988. In June 1988, a large illegal toxic waste dump which had been created by an Italian company was discovered. This led to the formation of the Basel Convention to stem the flow of poisonous substances from developed countries in 1989.

In 1991, the first electronic waste recycling system was implemented in Switzerland, beginning with collection of old refrigerators but gradually expanding to cover all devices. The organisation SWICO handles the programme, and is a partnership between IT retailers.

The first publication to report the recycling of computers and electronic waste was published on the front page of the New York Times on April 14, 1993 by columnist Steve Lohr. It detailed the work of Advanced Recovery Inc., a small recycler, in trying to safely dismantle computers, even if most waste was landfilled. Several other companies emerged in the early 1990s, chiefly in Europe, where national 'take back' laws compelled retailers to use them.

After these schemes were set up, many countries did not have the capacity to deal with the sheer quantity of e-waste they generated or its hazardous nature. They began to export the problem to developing countries without enforced environmental legislation. This is cheaper: the cost of recycling of computer monitors in the US is ten times more than in China. Demand in Asia for electronic waste began to grow when scrap yards found they could extract valuable substances such as copper, iron, silicon, nickel and gold, during the recycling process.

The Waste Electrical and Electronic Equipment Directive (WEEE Directive) became European Law in February 2003 and covers all aspects of recycling all types of appliance. This was followed by Electronic Waste Recycling Act, enshrined in Californian law in January 2005.

The 2000s saw a large increase in both the sale of electronic devices and their growth as a waste stream: in 2002 e-waste grew faster than any other type of waste in the EU. This caused investment in modern, automated facilities to cope with the influx of redundant appliances.

E-cycling

"E-cycling" or "E-waste" is an initiative by the United States Environmental Protection Agency (EPA) which refers to donations, reuse, shredding and general collection of used electronics. Generically, the term refers to the process of collecting, brokering, disassembling, repairing and recycling the components or metals contained in used or discarded electronic equipment, otherwise known as electronic waste (e-waste). "E-cyclable" items include, but are not limited to: televisions, computers, microwave ovens, vacuum cleaners, telephones and cellular phones, stereos, and VCRs and DVDs just about anything that has a cord, light or takes some kind of battery.

Investment in e-cycling facilities has been increasing recently due to technology's rapid rate of obsolescence, concern over improper methods, and opportunities for manufacturers to influence the secondary market (used and reused products). The higher metal prices is also having more recycling taking place. The controversy around methods stems from a lack of agreement over preferred outcomes.

World markets with lower disposable incomes, consider 75% repair and reuse to be valuable enough to justify 25% disposal. Debate and certification standards may be leading to better definitions, though civil law contracts, governing the expected process are still vital to any contracted process, as poorly defined as "e-cycling".

Pros of E-cycling

The e-waste disposal occurring after processing for reuse, repair of equipment, and recovery of metals may be unethical or illegal when e-scrap of many kinds is transported overseas to developing countries for such processing. It is transported as if to be repaired and/or recycled, but after processing the less valuable e-scrap becomes e-waste/pollution there. Another point of view is that the net environmental cost must be compared to and include the mining, refining and extraction with its waste and pollution cost of new products manufactured to replace secondary products which are routinely destroyed in wealthier nations, and which cannot economically be repaired in older or obsolete products. As an example of negative impacts of e-waste, pollution of groundwater has become so serious in areas surrounding China's landfills that water must be shipped in from 18 miles (29 km) away. However, mining of new metals can have even broader impacts on groundwater. Either thorough e-cycling processing, domestic processing or overseas repair, can help the environment by avoiding pollution. Such e-cycling can theoretically be a sustainable alternative to disposing of e-waste in landfills. In addition, e-cycling allows for the reclamation of potential conflict minerals, like gold and wolframite, which requires less of those to be mined and lessens the potential money flow to militias and other exploitative actors in third-world that profit from mining them.

Supporters of one form of "required e-cycling" legislation argue that e-cycling saves taxpayers money, as the financial responsibility would be shifted from the taxpayer to the manufacturers. Advocates of more simple legislation (such as landfill bans for e-waste) argue that involving manufacturers does not reduce the cost to consumers, if reuse value is lost, and the resulting costs are then passed on to consumers in new products, particularly affecting markets which can hardly afford new products. It is theorized that manufacturers who take part in e-cycling would be motivated to use fewer materials in the production process, create longer lasting products, and implement safer, more efficient recycling systems. This theory is sharply disputed and has never been demonstrated.

Criticisms of E-cycling

The critics of e-cycling are just as vocal as its advocates. According to the Reason Foundation, e-cycling only raises the product and waste management costs of e-waste for consumers and limits innovation on the part of high-tech companies. They also believe that e-cycling facilities could unintentionally cause great harm to the environment. Critics claim that e-waste doesn't occupy a significant portion of total waste. According to a European study, only 4% of waste is electronic.

Another opposition to e-cycling is that many problems are posed in disassembly: the process is costly and dangerous because of the heavy metals of which the electronic products are composed, and as little as 1-5% of the original cost of materials can be retrieved. A final problem that people find is that identity fraud is all too common in regards to the disposal of electronic products. As the programs are legislated, creating winners and losers among e-cyclers with different locations and processes, it may be difficult to distinguish between criticism of e-cycling as a practice, and criticism of the specific legislated means proposed to enhance it.

The Fate of E-waste

A hefty criticism often lobbed at reuse based recyclers is that people think that they are recycling their electronic waste, when in reality it is actually being exported to developing countries like China, India, and Nigeria. For instance, at free recycling drives, "recyclers" may not be staying true to their word, but selling e-waste overseas or to parts brokers. Studies indicate that 50-80% of the 300,000 to 400,000 tons (270,000 to 360,000 tonnes) of e-waste is being sent overseas, and that approximately 2 million tons (1.8 million tonnes) per year go to U.S. landfills.

Although not possible in all circumstances, the best way to e-cycle is to upcycle e-waste. On the other hand, the electronic products in question are generally manufactured, and repaired under warranty, in the same nations, which anti-reuse recyclers depict as primitive. Reuse-based e-recyclers believe that fair-trade incentives for export markets will lead to better results than domestic shredding. There has been a continued debate between export-friendly e-cycling and increased regulation of that practice.

In the European Union, debate regarding the export of e-waste has resulted in a significant amendment to the WEEE directive (January 2012) with a view to significantly diminishing the export of WEEE (untreated e-waste). During debate in Strasburg, MEPs stated that "53 million tonnes of WEEE were generated in 2009 but only 18% collected for recycling" with the remainder being exported or sent to landfill. The Amendment, voted through by a unanimous 95% of representatives, removed the re-use (repair and refurbishmet) aspect of the directive and placed more emphasis upon recycling and recovery of precious metals and base metals. The changes went further by placing the burden upon registered exporters to prove that used equipment leaving Europe was "fit for purpose".

Policy Issues and Current Efforts

Currently, pieces of government legislation and a number of grassroots efforts have contributed to the growth of e-cycling processes which emphasize decreased exports over increased reuse rates. The Electronic Waste Recycling Act was passed in California in 2003. It requires that consumers pay an extra fee for certain types of electronics, and the collected money be then redistributed to recycling companies that are qualified to properly recycle these products. It is the only state that legislates against e-waste through this kind of consumer fee; the other states' efforts focus on producer responsibility laws or waste disposal bans. No study has shown that per capita recovery is greater in one type of legislated program (e.g. California) versus ordinary waste disposal bans (e.g. Massachusetts), though recovery has greatly increased in states which use either method.

As of September, 2006, Dell developed the nation's first completely free recycling program, furthering the responsibilities that manufacturers are taking for e-cycling. Manufacturers and retailers such as Best Buy, Sony, and Samsung have also set up recycling programs. This program does not accept televisions, which are the most expensive used electronic item, and are unpopular in markets which must deal with televisions when the more valuable computers have been cherry picked.

Another step being taken is the recyclers' pledge of true stewardship, sponsored by the Computer TakeBack Campaign. It has been signed by numerous recyclers promising to recycle responsibly. Grassroots efforts have also played a big part in this issue, as they and other community organizations are being formed to help responsibly recycle e-waste. Other grassroots campaigns are Basel, the Computer TakeBack Campaign (co-coordinated by the Grassroots Recycling Network), and the Silicon Valley Toxics Coalition. No study has shown any difference in recycling methods under the Pledge, and no data is available to demonstrate difference in management between "Pledge" and non-Pledge companies, though it is assumed that the risk of making false claims will prevent Pledge companies from wrongly describing their processes.

Many people believe that the U.S. should follow the European Union model in regards to its management of e-waste. In this program, a directive forces manufacturers to take responsibility for e-cycling; it also demands manufacturers' mandatory take-back and places bans on exporting e-waste to developing countries. Another longer-term solution is for computers to be composed of less dangerous products and many people disagree. No data has been provided to show that people who agree with the European model have based their agreement on measured outcomes or experience-based scientific method.

Data Security

E-waste presents a potential security threat to individuals and exporting countries. Hard drives that are not properly erased before the computer is disposed of can be reopened, exposing sensitive information. Credit card numbers, private financial data, account information and records of online transactions can be accessed by most willing individuals. Organized criminals in Ghana commonly search the drives for information to use in local scams.

Electronic waste dump at Agbogbloshie, Ghana. Organized criminals commonly search the drives for information to use in local scams.

Government contracts have been discovered on hard drives found in Agbogbloshie, Ghana. Multimillion-dollar agreements from United States security institutions such as the Defense Intelligence Agency (DIA), the Transportation Security Administration and Homeland Security have all resurfaced in Agbogbloshie.

Reasons to Destroy and Recycle Securely

There are ways to ensure that not only hardware is destroyed but also the private data on the hard drive. Having customer data stolen, lost, or misplaced contributes to the ever growing number of people who are affected by identity theft, which can cause corporations to lose more than just money. The image of a company that holds secure data, such as banks, law firms, pharmaceuticals, and credit corporations is also at risk. If a company's public image is hurt, it could cause consumers to not use their services and could cost millions in business losses and positive public relation campaigns. The cost of data breaches "varies widely, ranging from $90 to $50,000 (under HIPAA's new HITECH amendment, that came about through the American Recovery and Revitalization act of 2009), as per customer record, depending on whether the breach is "low-profile" or "high-profile" and the company is in a non-regulated or highly regulated area, such as banking or medical institutions."

There is also a major backlash from the consumer if there is a data breach in a company that is supposed to be trusted to protect their private information. If an organization has any consumer info on file, they must by law (Red Flags Clarification act of 2010) have written information protection policies and procedures in place, that serve to combat, mitigate, and detect vulnerable areas that could result in identity theft. The United States Department of Defense has published a standard to which recyclers and individuals may meet in order to satisfy HIPAA requirements.

Secure Recycling

Countries have developed standards, aimed at businesses and with the purpose of ensuring the security of Data contained in 'confidential' computer media [NIST 800-88: US standard for Data Remenance][HMG CESG IS5, Baseline & Enhanced, UK Government Protocol for Data Destruction]. National Association for Information Destruction (NAID) "is the international trade association for companies providing information destruction services. Suppliers of products, equipment and services to destruction companies are also eligible for membership. NAID's mission is to promote the information destruction industry and the standards and ethics of its member companies." There are companies that follow the guidelines from NAID and also meet all Federal EPA and local DEP regulations.

The typical process for computer recycling aims to securely destroy hard drives while still recycling the byproduct. A typical process for effective computer recycling:

1. Receive hardware for destruction in locked and securely transported vehicles.
2. Shred hard drives.
3. Separate all aluminum from the waste metals with an electromagnet.
4. Collect and securely deliver the shredded remains to an aluminum recycling plant.
5. Mold the remaining hard drive parts into aluminum ingots.

The Asset Disposal and Information Security Alliance (ADISA) publishes an *ADISA IT Asset Disposal Security Standard* that covers all phases of the e-waste disposal process from collection to transportation, storage and sanitization's at the disposal facility. It also conducts periodic audits of disposal vendors.

Ship Breaking

Ship breaking or ship demolition is a type of ship disposal involving the breaking up of ships for either a source of parts, which can be sold for re-use, or for the extraction of raw materials, chiefly scrap. It may also be known as ship dismantling, ship cracking, or ship recycling. Modern ships have a lifespan of 25 to 30 years before corrosion, metal fatigue and a lack of parts render them uneconomical to run. Ship breaking allows the materials from the ship, especially steel, to be recycled and made into new products.

This lowers the demand for mined iron ore and reduces energy use in the steel-making process. Equipment on board the vessel can also be reused. While ship breaking is, in theory, sustainable, there are concerns about the use of poorer countries without stringent environmental legislation. It is also considered one of the world's most dangerous industries and very labour-intensive.

Workers drag steel plate ashore from beached ships in Chittagong, Bangladesh.

In 2012, roughly 1,250 ocean ships were broken down, and their average age was 26 years. In 2013, the world total of demolished ships amounted to 29,052,000 tonnes, 92% of which were demolished in Asia. India, Bangladesh, China and Pakistan have the highest market share and are global centres of ship breaking, with Alang in India and Gadani in Pakistan being the largest ships graveyards in the world. The largest sources of ships are states of China, Greece and Germany respectively, although there is a greater variation in the source of carriers versus their disposal. The ship breaking yards of India, Bangladesh, China and Pakistan employ 100,000 workers as well as providing a large amount of indirect jobs. In India, the recycled steel covers 10% of the country's needs.

As an alternative to ship breaking, ships may be sunk to create artificial reefs after being cleared of hazardous materials, or sunk in deep ocean waters. Storage is a viable temporary option, whether on land or afloat, though all ships will be eventually scrapped, sunk, or preserved for museums.

History

Wooden-walled ships were simply set on fire or 'conveniently sunk'. In Tudor times, ships were also dismantled and the timber re-used. This procedure was no longer applicable with the advent of metal-hulled boats.

There is evidence of the industry as early as 1838. The navy vessel HMS Temeraire had her masts, stores and guns removed and her crew paid off. She was sold by Dutch auction on 16 August 1838 to John Beatson, a shipbreaker based at Rotherhithe for £5,530. Beatson was then faced with the task of transporting the ship 55 miles from Sheerness to Rotherhithe, the largest ship to have attempted this voyage. To accomplish this he hired two steam tugs from the Thames Steam Towing Company and employed a Rotherhithe pilot named William Scott and twenty five men to sail her up the Thames, at a cost of £58. The shipbreakers undertook a thorough dismantling, removing all the copper sheathing, rudder pintles and gudgeons, copper bolts, nails and other fastenings to be sold back to the Admiralty. The timber was mostly sold to house builders and shipyard owners, though some was retained for working into specialist commemorative furniture. The ship's final voyage was immortalised by William Turner's painting The Fighting 'Temeraire', Tugged to her Last Berth to be Broken Up, 1838.

HMS *Queen* heeled over on the Thames foreshore off Rotherhithe.

In 1880, Denny Brothers of Dumbarton used scrap maritime steel in their shipbuilding. Many other nations began to purchase British ships for scrap by the late 19th century, including Germany, Italy, the Netherlands and Japan. The Italian industry started in 1892, and the Japanese after an 1896 law had been passed to subsidise native shipbuilding.

After being damaged or involved in a disaster, liner operators did not want the name of the broken ship to tarnish the brand of their passenger services. The final voyage of many Victorian ships was with the final letter of their name chipped off.

The armistice of 1918 brought a glut of naval 'men-o-war' into the ship breaking industry which boomed and subsequently died down in the early 1920s. In the 1930s, it became cheaper to 'beach' a boat and run her ashore as opposed to using a dry dock. The ship would have to weigh as little as possible and run ashore at full speed. Dismantling operations required a 10 feet rise of tide and close proximity to a steel-works. Electric shears, a wrecking ball and oxy-acetylene torches were used. The technique of the time is almost identical to that of developing countries today. Similarly, Thos W Ward Ltd., one of the largest breakers in the United Kingdom in the 1930s, would recondition and sell all furniture and machinery. Many historical artefacts were sold at public auctions: the Cunarder Mauretania received high bids for her fittings worldwide. However, even with obsolete technology, any weapons and military information were carefully removed.

Until the late 20th century, ship breaking took place in port cities of industrialized countries such as the United Kingdom and the United States. Those dismantlers that still remain in the United States work primarily on government surplus vessels. In the mid 20th century, low-cost East Asian countries began to dominate ship breaking, with countries such as Japan, then Korea and Taiwan and then China increasing their world share. For example, in 1977 Taiwan dominated the industry with more than half the market share, followed by Spain and Pakistan. Bangladesh had no capacity at all. However, the sector is volatile and fluctuates wildly, and Taiwan processed just 2 ships 13 years later as wages across East Asia rose.

Dismantling of *Redoutable* in Toulon, 1912.

In 1960, after a severe cyclone, the Greek ship *M D Alpine* was stranded on the shores of Sitakunda, Chittagong. It could not be re-floated and so remained there for several years. In 1965, the then in East Pakistan, Chittagong Steel House bought the ship and had it scrapped. It took years to scrap the vessel, but the work gave birth to the industry in Bangladesh. Until 1980 Gadani ship-breaking yard of Pakistan was the largest ship-breaking yard of the world.

Tightening environmental regulations resulted in increased costs of hazardous waste disposal in industrialised countries in the 1980s, causing ships to be exported to lower income nations, chiefly South Asia. This, in turn, created a far worse environmental problem, subsequently leading to the Basel Convention. In 2004 a Basel Convention decision officially classified old ships as "toxic waste", preventing them from leaving a country without the permission of the importing state. This has led to a resurgence of recycling in environmentally-compliant locations in developed countries, especially in former ship building yards.

On 31 December 2005, the French Navy's *Clemenceau* left Toulon to be dismantled in Alang, India despite protests over improper disposal capabilities and facilities for the toxic wastes. On 6 January 2006 the Supreme Court of India temporarily denied access to Alang, and the *Conseil d'État* ordered *Clemenceau* to return to French waters. Able UK in Hartlepool received a new disassembly contract to use accepted practices in scrapping the ship. The dismantling started on 18 November 2009 and the break-up was completed by the end of 2010, and the event was considered a turning point in the treatment of redundant vessels. Europe and the United States have actually had a resurgence in ship scrapping since the 1990s.

In 2009 the Bangladesh Environmental Lawyers Association won a legal case prohibiting all substandard ship breaking. For 14 months the industry could not import ships and thousands of jobs were lost before the ban was annulled. That same year, the global recession and lower demand for goods led to an increase in the supply of ships for decommissioning. The rate of scrapping is inversely correlated to the freight price, which collapsed in 2009.

Technique

The decommissioning process is entirely different in developed and developing countries. Both start with an auction for which the highest bidder wins the contract. The ship-breaker then acquires the vessel from the international broker who deals in outdated ships. The price paid is around $400 per tonne ($4–10 million), and the poorer the environmental legislation the higher the price. The purchase of water-craft makes up 69% of the income earned by the industry in Bangladesh, versus 2% for labour costs. The boat is taken to the decommissioning location either under its own power or with the use of tugs.

An aerial view of Chittagong, Bangladesh.

Developing Countries

In developing countries, chiefly the Indian subcontinent, ships are run ashore on gently sloping sand tidal beaches at high tide so that they can be accessed for disassembly. As described in *"History"* (above), the sizeable ship breaking industry of Bangladesh traces its origin to a ship beached there accidentally during a cyclone. Manoeuvring a large ship onto a beach at high speed takes skill and daring even for a specialist captain, and is not always successful. Next, the anchor is dropped to steady the ship and the engine is shut down. It takes 50 labourers about three months to break down a normal-sized cargo vessel of about 40,000 tonnes.

The decommissioning begins with the draining of fuel and fire fighting liquid, which is sold to the trade. Any re-usable items—wiring, furniture and machinery—are sent to local markets or the trade. Unwanted materials become inputs to their relevant waste streams. Often, in lower income nations, these industries are no better than ship breaking. For example, the toxic insulation is usually burnt off copper wire to access the metal. Some crude safety precautions exist—chickens are lowered into the chambers of the ship, and if the birds return alive, they are considered safe. Workers also do not have proper clothing, footwear and masks.

Gas cutting in Chittagong, Bangladesh.

Sledgehammers and oxy-acetylene gas-torches are used to cut up the steel hull. Cranes are not typically used on the ship, because of costs. Pieces of the hull simply fall off and are dragged inland, possibly aided with a winch or bulldozer. These are then cut into smaller pieces away from the coast. 90% of the steel is re-rollable scrap: higher quality steel plates that are heated and reused as reinforcement bar for construction. The remainder is transported to electric arc furnaces to be melted down into ingots for re-rolling mills. In the re-rolling mills, the heating of painted steel plates (in particular, those painted with chlorinated rubber paints) generates dioxins. Substances which are costly to dispose of, such as hazardous waste, are left on the beach or set on fire, even old batteries and half-empty cans of paint. Stockpiled in Bangladesh, for example, are 79,000 tonnes of asbestos, 240,000 tonnes of PCBs and 210,000 tonnes of Ozone-depleting substances.

Developed Countries

In developed countries the dismantling process should mirror the technical guidelines for the environmentally sound management of the full and partial dismantling of ships, published by the Basel Convention in 2003. Recycling rates of 98% can be achieved in these facilities.

Prior to dismantling, an inventory of dangerous substances should be compiled. All hazardous materials and liquids, such as bilge water, are removed before ship breaking. Holes should be bored for ventilation and all flammable vapours are extracted.

Vessels are initially taken to a dry dock or a pier, although a dry dock is considered more environmentally friendly because all spillage is contained and can easily be cleaned up.

Floating is, however, cheaper than a dry dock. Storm water discharge facilities will stop an overflow of toxic liquid into the waterways. The carrier is then secured to ensure its stability. Often the propeller is removed beforehand to allow the water-craft to be moved into shallower water.

Workers must completely strip the ship down to a bare hull, with objects cut free using saws, grinders, abrasive cutting wheels, hand held shears, plasma and gas torches. Anything of value, such as spare parts and electronic equipment is sold for re-use, although labour costs mean that low value items are not economical to sell. The Basel Convention demands that all yards separate hazardous and non-hazardous waste and have appropriate storage units, and this must be done before the hull is cut up. Asbestos, found in the engine room, is isolated and stored in custom-made plastic wrapping prior to being placed in secure steel containers, which are then landfilled.

Many hazardous wastes can be recycled into new products. Examples include lead-acid batteries or electronic circuit boards. Another commonly used treatment is cement based solidification and stabilization. Cement kilns are used because they can treat a range of hazardous wastes by improving physical characteristics and decreasing the toxicity and transmission of contaminants. A hazardous waste may also be "destroyed" by incinerating it at a high temperature; flammable wastes can sometimes be burned as energy sources. Some hazardous waste types may be eliminated using pyrolysis in a high temperature electrical arc, in inert conditions to avoid combustion. This treatment method may be preferable to high temperature incineration in some circumstances such as in the destruction of concentrated organic waste types, including PCBs, pesticides and other persistent organic pollutants. Dangerous chemicals can also be permanently stored in landfills as long as leaching is prevented.

Valuable metals, such as copper in electric cable, that are mixed with other materials may be recovered by the use of shredders and separators in the same fashion as e-waste recycling. The shredders cut the electronics into metallic and non-metallic pieces. Metals are extracted using magnetic separators, air flotation separator columns, shaker tables or eddy currents. The plastic almost always contains regulated hazardous waste (e.g., asbestos, PCBs, hydrocarbons) and cannot be melted down.

Large objects, such as engine parts, are extracted and sold as they become accessible. The hull is cut into 300 tonne sections, starting with the upper deck and working slowly downwards. While oxy-acetylene gas-torches are most commonly used, detonation charges can quickly remove large sections of the hull. These sections are transported to an electric arc furnace to be melted down into new ferrous products, though toxic paint must be stripped prior to heating.

Health Risks

70% of ships are simply run ashore in developing countries for disassembly, where (particularly in older vessels) asbestos, lead, polychlorinated biphenyls and heavy metals pose a danger for the workers. Burns from explosions and fire; suffocation; mutilation from falling metal; cancer and disease from toxins are regular occurrences in the industry. Asbestos was used heavily in ship construction until it was finally banned in most of the developed world in the mid-1980s. Currently, the costs associated with removing asbestos, along with the potentially expensive insurance and health risks, have meant that ship breaking in most developed countries is no longer economically viable. Dangerous vapors and fumes from burning materials can be inhaled, and dusty asbestos-laden areas are commonplace.

Removing the metal for scrap can potentially cost more than the value of the scrap metal itself. In the developing world, however, shipyards can operate without the risk of personal injury lawsuits or workers' health claims, meaning many of these shipyards may operate with high health risks. Protective equipment is sometimes absent or inadequate. The sandy beaches cannot sufficiently support the heavy equipment, which is thus prone to collapse. Many are injured from explosions when flammable gas is not removed from fuel tanks. In Bangladesh, a local watchdog group claims that one worker dies a week and one is injured per a day on average.

The problem is caused by negligence from national governments, shipyard operators, and former ship owners disregarding the Basel Convention. According to the Institute for Global Labour and Human Rights, workers who attempt to unionize are fired and then blacklisted. The employees have no formal contract or any rights, and sleep in over-crowded hostels. The authorities produce no comprehensive injury statistics, so the problem is underestimated. Child labour is also widespread: 20% of Bangladesh's ship breaking workforce are below 15 years of age, mainly involving in cutting with gas torches. Greenpeace discovered that the yard owners and local authorities are complicit in attempting to cover up the problem in order to reap the economic benefits.

There is, however, an active ship breaker's union in Mumbai, India (Mumbai Port Trust Dock and General Employees' Union) since 2003 with 15,000 members, which strikes to ensure fatality compensation. It has set up a sister branch in Alang, gaining paid holidays and safety equipment for workers since 2005. They hope to expand all along the South Asian coastline.

Several United Nations committees are increasing their coverage of ship breakers' human rights. In 2006, the International Maritime Organisation developed legally binding global legislation which concerns vessel design, vessel recycling and the enforcement of regulation thereof and a 'Green Passport' scheme. Water-craft must have an inventory of hazardous material before they are scrapped, and the facilities must meet health & safety requirements. The International Labour Organization created a voluntary set of guidelines for occupational safety in 2003. Nevertheless, Greenpeace found that even pre-existing mandatory regulation has had little noticeable effect for labourers, due to government corruption, yard owner secrecy and a lack of interest from countries who prioritise economic growth. To safeguard worker health, the report recommends that developed countries create a fund to support their families, certify carriers as 'gas-free' (i.e. safe for cutting) and to remove toxic materials in appropriate facilities before export. To supplement the international treaties, organisations such as the NGO Shipbreaking Platform, the Institute for Global Labour and Human Rights and ToxicsWatch Alliance are lobbying for improvements in the industry. There are also guards who look out for any reporters.

Environmental Risks

In recent years, ship breaking has become an issue of environmental concern beyond the health of the yard workers. Many ship breaking yards operate in developing nations with lax or no environmental law, enabling large quantities of highly toxic materials to escape into the general environment and causing serious health problems among ship breakers, the local population, and wildlife. Environmental campaign groups such as Greenpeace have made the issue a high priority for their activities.

Along the Indian subcontinent, ecologically-important mangrove forests, a valuable source of protection from tropical storms and monsoons, have been cut down to provide space for water-craft disassembly. In Bangladesh, for example, 40,000 mangrove trees were illegally chopped down in 2009. The World Bank has found that the country's beaching locations are now at risk from sea level rise. 21 fish and crustacean species have been wiped out in the country as a result of the industry as well. Lead, organotins such as tributyltin in anti-fouling paints, polychlorinated organic compounds, by-products of combustion such as polycyclic aromatic hydrocarbons, dioxins and furans are found in ships and pose a great danger to the environment.

The Basel Convention on the Control of Trans-boundary Movements of Hazardous Wastes and Their Disposal of 1989 has been ratified by 166 countries, including India and Bangladesh, and in 2004, End of Life Ships were subjected to its regulations. It aims to stop the transportation of dangerous substances to less developed countries and mandate the use of regulated facilities. However, Greenpeace reports that neither vessel exporter nor breaking countries are adhering to its policies. The organisation recommends that all parties enforce the Basel Convention in full, and hold those who break it liable. Furthermore, the decision to scrap a ship is often made in international waters, where the convention has no jurisdiction.

The Hong Kong Convention is a compromise. It allows ships to be exported for recycling, as long as various stipulations are met: All water-craft must have an inventory and every shipyard needs to publish a recycling plan to protect the environment. The Hong Kong Convention was adopted in 2009 but with few countries signing the agreement.

In March 2012 the European Commission proposed tougher regulations to ensure all parties take responsibility. Under these rules, if a vessel has a European flag, it must be disposed of in a shipyard on an EU "green list." The facilities would have to show that they are compliant, and it would be regulated internationally in order to bypass corrupt local authorities. However, there is evidence of ship owners changing the flag to evade the regulations. China's scrap industry has vehemently protested against the proposed European regulations. Although Chinese recycling businesses are less damaging than their South Asian counterparts, European and American ship breakers comply with far more stringent legislation.

List of Ship Breaking Yards

The following are some of world's largest ship breaking yards:

Bangladesh

- Chittagong Ship Breaking yard

China

- Changjiang Ship Breaking yard, located in Jiangyin, China

India

- Alang-Sosiya Ship Breaking Yard

Pakistan

- Gadani Ship Breaking yard

Turkey

- Aliaga Ship Breaking Yard

United States

- Esco Marine, Brownsville, Texas
- International Shipbreaking, Brownsville, Texas

United Kingdom

- Able UK, Graythorpe Dock, Teeside

Belgium

- Galloo, Ghent, formerly Van Heyghen Recycling

Ferrous Metal Recycling

Ferrous metals are able to be recycled, with steel being one of the most recycled materials in the world,. Ferrous metals contain an appreciable percentage of iron and the addition of carbon and other substances creates steel.

A pile of steel scrap in Brussels, waiting to be recycled.

Description

In the United States, steel containers, cans, automobiles, appliances, and construction materials contribute the greatest weight of recycled materials. For example, in 2008, more than 97% of structural steel and 106% of automobiles were recycled, comparing the current steel consumption for each industry with the amount of recycled steel being produced (the late 2000s recession and the associated sharp decline in automobile production in the US explains the over-100% calculation). A typical appliance is about 75% steel by weight and automobiles are about 65% steel and iron.

The Universal Symbol for Recyclable Steel.

The steel industry has been actively recycling for more than 150 years, in large part because it is economically advantageous to do so. It is cheaper to recycle steel than to mine iron ore and manipulate it through the production process to form new steel. Steel does not lose any of its inherent physical properties during the recycling process, and has drastically reduced energy and material requirements compared with refinement from iron ore. The energy saved by recycling reduces the annual energy consumption of the industry by about 75%, which is enough to power eighteen million homes for one year. According to the International Resource Panel's Metal Stocks in Society report, the per capita stock of steel in use in Australia, Canada, the European Union EU15, Norway, Switzerland, Japan, New Zealand and the US combined is 7,085 kilograms (15,620 lb) (about 860 million people in 2005).

The CEN Symbol for Recyclable Steel.

Basic oxygen steelmaking (BOS) uses 25–35% recycled steel to make new steel. BOS steel usually contains lower concentrations of residual elements such as copper, nickel and molybdenum and is therefore more malleable than electric arc furnace (EAF) steel and is often used to make automotive fenders, tin cans, industrial drums or any product with a large degree of cold working. EAF steelmaking uses almost 100% recycled steel. This steel contains greater concentrations of residual elements that cannot be removed through the application of oxygen and lime. It is used to make structural beams, plates, reinforcing bar and other products that require little cold working. Downcycling of steel by hard-to-separate impurities such as copper or tin can only be prevented by well-aimed scrap selection or dilution by pure steel. Recycling one metric ton (1,000 kilograms) of steel saves 1.1 metric tons of iron ore, 630 kilograms of coal, and 55 kilograms of limestone.

Types of Scrap Used in Steelmaking

- Heavy melting steel – Industrial or commercial scrap steel greater than 6mm thick, such as plates, beams, columns, channels; may also include scrap machinery or implements or certain metal stampings
- Old car bodies – Vehicles with or without interiors and their original wheels
- Cast iron – Cast iron bathtubs, machinery, pipe and engine blocks
- Pressing steel – Domestic scrap metal up to approx. 6mm thick. Examples - White goods (fridges, washing machines, etc.), roofing iron, water heaters, water tanks and sheet metal offcuts
- Re-inforcing bars or mesh – Used in the construction industry within concrete structures
- Turnings – Remains of drilling or shaping steels. Also known as “borings” or “swarf”
- Manganese steel – Non magnetic, hardened steel used in the mining industry, cement mixers, rock crushers, and other high impact and abrasive environments
- Rails – Rail or tram tracks

Recycling by Country

United States

As of 2008, more than 83% of steel was recycled in the United States. It is the most widely recycled material; in 2000, more than 60 million metric tons were recycled.

Precycling

Precycling is the practice of reducing waste by attempting to avoid bringing items which will generate waste into home or business. The U.S. Environmental Protection Agency (EPA) also cites that precycling is the preferred method of integrated solid waste management because it cuts waste at its source and therefore trash is eliminated before it is created. According to the EPA, precycling is also characterized as a decision-making process on the behalf of the consumer because it involves making informed judgments regarding a product's waste implications. The implications that are taken into consideration by the consumer include: whether a product is reusable, durable, or repairable; made from renewable or non-renewable resources; over-packaged; and whether or not the container is reusable.

About

Precycling has the ability to build industrial, social, environmental, and economic circumstances that allow for old products to be converted into new resources;

- Industrial: increasing the independence from accumulative substances, such as heavy metals, fossil fuels, synthetics, etc.
- Economic: create a circular economy.
- Ecological/environmental: allowance for more extensive and diverse natural habitats where the resources are returned to nature.
- Societal: extend the capacity of precycling to meet everyone's needs.

The concept of 'precycling' was coined in 1988 by social marketing executive Maureen O'Rorke in a public waste education campaign for the City of Berkeley. The application of precycling is not limited to large corporations, but can be administered on smaller scales in local communities. The reason precycling is effective on large scales and on small scales stems from the idea that it shares a common language between experts and non-experts, buyers and sellers, economists and environmentalists. However, it is important to consider that waste prevention systems, such as precycling, require the collaborative effort from several working parts. These parts include prevention targets, producer responsibility, householder charging, funding for pilot projects, public involvement, engagement of private and third sectors, and public campaigns that spread awareness.

Integration of Waste Management

The original three-pronged push for trash management is “Reduce, Reuse, Recycle.” Precycling emphasizes “reducing and reusing”, while harnessing and questioning the momentum and popularity of the term “recycle.” In addition to this three-pronged waste management strategy of “Reduce, Reuse, Recycle”, precycling incorporates four supplementary R’s: Repair, Recondition, Remanufacture and Refuse. Waste is a resource that can be reused, recycled, recovered, or treated. Precycling differs from other singular forms of waste prevention because it encompasses not one behavior, but many.

Reduce

Reduce is a form of precycling that allows for the preservation of natural resources and also saves money on behalf of the manufacturer, the consumer, and the waste manager. Moreover, effective source reduction slows the depletion of environmental resources, prolongs the life of waste management facilities, and makes combustion and landfills safer by removing toxic waste components.

Reuse

Reuse is a form of precycling that reinvents items after their initial life and avoids creating additional waste.

Recycle

Although precycling harnesses the familiarity of the term recycling, it is still important to note the difference between recycling and prevention. Since precycling focuses on the prevention of waste production, this entails that measures are taken before a substance, material, or product has become waste. Whereas recycling is a type of precycling that involves taking action before existing waste is abandoned to nature. Recycling is a process where discarded materials are collected, sorted, processed, and used in the production of new products. Every time a person engages in the act of recycling, they help increase the market and bring the cost down. However, current research from the American Plastics Council states that only 25% of the nation’s recycling capabilities are being utilized.

Traditionally recycling requires large amounts of energy to “melt down” and then re-manufacture items. While this may cut down on the amount of trash that is going into landfills, it is not sustainable unless the underlying energy supply is sustainable. In addition, recycling often means downcycling and always involves at least some loss of the original material, so primary extraction is still required to make up the difference. Precycling reduces these problems by using less material in the first place, so less has to be recycled.

Repairing

Repair is a type of precycling that corrects specified faults in a product, however the quality of a repaired product is inferior to reconditioned or remanufactured items. One survey found that 68% of the respondents believed repairing was not cost efficient and sought alternative methods such as reconditioning or remanufacturing.

Reconditioning

Reconditioning is a type of precycling that requires the rebuilding of major components to restore a product's working condition, which is expected to be inferior to the original product.

Remanufacturing

Remanufacturing is another type of precycling that involves the greatest degree of work content, which results in superior quality of the product. In order to remanufacture a product, it requires a total dismantling of the product and the subsequent restoration and replacement of its parts. Remanufacturing is a preferred method of waste reduction compared to repairing and reconditioning because it preserves the embodied energy that has been used to shape the components of a product for their first life and it only requires 20-25% of the initial energy used in formation.

Refuse

Refusal to buy certain products due to detrimental impacts on the environment or wasteful packaging is another type of precycling because the rejection of such items paves the way for products that can be reduced, reused, or recycled.

Zero-waste Strategy

A zero waste approach aims to prevent rather than just reduce accumulated waste. Zero-waste goes beyond recycling to include the whole system, which includes the flow of resources and waste through human society. This "design principle" works to maximize recycling, minimize waste, reduce consumption and ensures that products are reused, repaired or recycled back into nature or the market. This preventative approach is more manageable and effective than incremental approaches that focus on gradually reducing the amount of impact because it is less complex and contains less information, which permits wider public participation.

Sustainability

In regards to sustainability, the term itself is often associated with resource constraints and maintenance of the status quo rather than growth and prosperity. However, with the implementation of a zero-waste management strategy, sustainable practices can push the status quo in order to create a society that is capable of development, technically and culturally advanced, dynamic in population and production, thoughtful with the use of non-renewable resources, and diverse, democratic, and challenging.

Economic Effects

Increased waste production is often negatively associated with increased economic growth. However, a zero-waste management strategy allows for economic growth that works cohesively with sustainability rather than against it. The implementation of a zero-waste strategy is part of an economic goal-set that aims to create a circular economy. A circular economy refers to a closed-loop socio-economic system that focuses on minimizing wastes while simultaneously maximizing stocks of resources for the economy. This closed-loop design diverts linear (open-loop) waste disposal streams into new raw material streams.

In a circular economy, one way to minimize waste is through the employment of precycling insurance, which allows for a full range of financed waste prevention opportunities. This type of insurance would set premiums related to the risk of a product ending up as waste, and these premiums would serve to fund actions concerning waste prevention. When establishing a premium for precycling insurance several factors need to calculated: recyclability or biodegradability; provision of infrastructure, habitat or collaborations for the generation of the product from new resources; the ecosystem concentrations of product components above natural levels. The idea of precycling insurance is plausible considering the aim of insurance industries is to avoid losses rather than paying for losses. However, in order for this idea to work, private and third sectors need to be involved and engaged in the issue. In this instance, a third sector refers to small charities and a handful of societal enterprises that coordinate with charity shops.

Environmental Effects

According to the "Extended Producer Responsibility" principle, impacts are substantially determined at the point of design where key decisions are made on materials, production process, and how products are used and disposed of at the end of life-cycle, which falls on the producer. However, in a circular economy there is the recognition that nature's capacity needs to be maximized through the reprocess of biodegradable wastes produced by industries and human activity. This task is accomplished through the procurement and funding of precycling insurance premiums that invest in systematic preservation of endangered habitats, careful harvesting of biological resources and expansions of productive ecosystems. Additionally, in terms of climate change, precycling insurance offers a flexible alternative to the binding limits on greenhouse gas emissions and international taxation on mineral fuels. In terms of waste management systems, the environment benefits from the reparation of products to the greatest degree because less energy is required and the majority of the original material is kept intact.

Societal Effects

The social structure operating under a circular economy is referred to as a circular society. The aim of a circular society is to create a cooperative culture by means of problem-prevention, resource-availability and fuller participation, with reference to precycling. One critique of this approach, in terms of waste management, is that It is difficult to maintain a cooperative culture within a society because they it is constantly evolving and changing.

Raising Awareness

There is an increasing public awareness on the need for sustainable production and consumption. One campaign that aimed at raising awareness of precycling focused on whether people's self-reported behaviors were affected by exposure to precycling advertisements on the radio, television, or in-store flyers. The researchers concluded that the most effective results stemmed from the inclusion of social rewards that invoke an intrinsic motivation to engage in precycling behaviors.

Another way to raise awareness is through statistics that highlight the potential impacts that can be achieved through waste prevention. For instance, if 70 million Americans bought a half-gallon container of milk each week (instead of two quarts), then 41.6 million pounds of paper discards and 5.7 pounds of plastic discards would be reduced annually. This transition from two quarts to a half-gallon would save $145.6 million on packaging each year.

Implementation

In order to effectively implement precycling practices and behaviors, the public needs to feel "enabled", "engaged", "encouraged", and "exemplified" in their efforts to partake in precycling.

Not only can the average consumer practice precycling, but industries can also participate. Purchasing from parts suppliers, reuse of chemicals, and reduction of unnecessary packaging are some methods. There are some companies and countries that have taken it upon themselves to implement more sustainable practices that align with precycling principles. For instance, Fonterra reduced its packaging through the implementation of bulking, reuse and redesign. Further, Waste Management New Zealand created Recycle New Zealand, which provided a subsidiary focusing on the collection of materials that could be diverted and sorted prior to the operations of reducing, recycling, or recovering. Moving forward, free-trade organizations can further implement precycling practices by exploring this strategy as a new way to reduce regulations and to promote greater industrial freedom of choice.

Moreover, the individual consumer can develop precycling habits by engaging in the following practices and behaviors:

"Enviro-shopping"

Enviro-shopping is considered shopping with the environment and implements a precycling strategy:

- Bringing one's own grocery bag or bring old ones back to the store
- Buying packages with the least amount of packaging
- Buying in bulk, but not buying more than one will use
- Looking for products with reusable dishes

Product Selection

Products to choose from in accordance with precycling principles:

- Plastic milk jugs or glass milk containers (no cartons)
- Fresh fruit and vegetables
- Concentrated products that involves less packaging
- Recycled products
- Rechargeable batteries

Behaviors

In addition to shopping practices that implement precycling principles, there are also behaviors that can be undertaken to prevent waste:

- Home composting
- Avoid junk mail
- Buy second-hand
- One way to participate in precycling is to carry a "precycling kit". Include a Tupperware or non-disposable container, silverware set, a cloth napkin or handkerchief, and a thermos or water-bottle within a cloth bag that can double as a grocery/shopping bag.

References

- Nguemaleu, Raoul-Abelin Choumin; Montheu, Lionel (2014-05-09). Roadmap to Greener Computing. CRC Press. p. 170. ISBN 9781466506848
- "Report: "On the Making of Silk Purses from Sows' Ears," 1921: Exhibits: Institute Archives & Special Collections: MIT". mit.edu. Retrieved 17, July 2020
- Lynn R. Kahle; Eda Gurel-Atay, eds. (2014). Communicating Sustainability for the Green Economy. New York: M.E. Sharpe. ISBN 978-0-7656-3680-5
- Huesemann, M.; Huesemann, J. (2011). Techno-fix: Why Technology Won't Save Us or the Environment. New Society Publishers. p. 464. ISBN 978-0-86571-704-6. Retrieved 16, May 2020
- Goodman, Peter S. (11 Jan 2012). "Where Gadgets Go To Die: E-Waste Recycler Opens New Plant In Las Vegas". The Huffington Post. Retrieved 25, June 2020
- Carl A. Zimring (2005). Cash for Your Trash: Scrap Recycling in America. New Brunswick, NJ: Rutgers University Press. ISBN 0-8135-4694-X
- The Self-Sufficiency Handbook: A Complete Guide to Greener Living by Alan Bridgewater pg. 62--Skyhorse Publishing Inc., 2007 ISBN 1-60239-163-7, ISBN 978-1-60239-163-5
- Sahni, S.; Gutowski, T. G. (2011). "Your scrap, my scrap! The flow of scrap materials through international trade" (PDF). IEEE International Symposium on Sustainable Systems and Technology (ISSST): 1–6. doi:10.1109/ISSST.2011.5936853. ISBN 978-1-61284-394-0
- "Bulgaria opens largest WEEE recycling factory in Eastern Europe". www.ask-eu.com. WtERT Germany GmbH. 12 Jul 2010. Retrieved 13, July 2020
- Cleveland, Cutler J.; Morris, Christopher G. (November 15, 2013). Handbook of Energy: Chronologies, Top Ten Lists, and Word Clouds. Elsevier. p. 461. ISBN 978-0-12-417019-3

Permissions

All chapters in this book are published with permission under the Creative Commons Attribution Share Alike License or equivalent. Every chapter published in this book has been scrutinized by our experts. Their significance has been extensively debated. The topics covered herein carry significant information for a comprehensive understanding. They may even be implemented as practical applications or may be referred to as a beginning point for further studies.

We would like to thank the editorial team for lending their expertise to make the book truly unique. They have played a crucial role in the development of this book. Without their invaluable contributions this book wouldn't have been possible. They have made vital efforts to compile up to date information on the varied aspects of this subject to make this book a valuable addition to the collection of many professionals and students.

This book was conceptualized with the vision of imparting up-to-date and integrated information in this field. To ensure the same, a matchless editorial board was set up. Every individual on the board went through rigorous rounds of assessment to prove their worth. After which they invested a large part of their time researching and compiling the most relevant data for our readers.

The editorial board has been involved in producing this book since its inception. They have spent rigorous hours researching and exploring the diverse topics which have resulted in the successful publishing of this book. They have passed on their knowledge of decades through this book. To expedite this challenging task, the publisher supported the team at every step. A small team of assistant editors was also appointed to further simplify the editing procedure and attain best results for the readers.

Apart from the editorial board, the designing team has also invested a significant amount of their time in understanding the subject and creating the most relevant covers. They scrutinized every image to scout for the most suitable representation of the subject and create an appropriate cover for the book.

The publishing team has been an ardent support to the editorial, designing and production team. Their endless efforts to recruit the best for this project, has resulted in the accomplishment of this book. They are a veteran in the field of academics and their pool of knowledge is as vast as their experience in printing. Their expertise and guidance has proved useful at every step. Their uncompromising quality standards have made this book an exceptional effort. Their encouragement from time to time has been an inspiration for everyone.

The publisher and the editorial board hope that this book will prove to be a valuable piece of knowledge for students, practitioners and scholars across the globe.

Index

www.ingramcontent.com/pod-product-compliance
Lightning Source LLC
LaVergne TN
LVHW080727210926
848485LV00003B/530
* 9 7 8 1 6 3 9 8 9 2 4 2 6 *